Floriculture: Principles and Practices

Floriculture: Principles and Practices

Violet Dalton

R CALLISTO REFERENCE

www.callistoreference.com

Callisto Reference,
118-35 Queens Blvd., Suite 400,
Forest Hills, NY 11375, USA

Visit us on the World Wide Web at:
www.callistoreference.com

ISBN: 978-1-64116-531-0 (Hardback)

Cataloging-in-Publication Data

Floriculture : principles and practices / Violet Dalton.
    p. cm.
Includes bibliographical references and index.
ISBN 978-1-64116-531-0
1. Floriculture. 2. Ornamental horticulture. 3. Flower gardening. I. Dalton, Violet.
SB404.9 .F56 2022
635.9--dc23

Table of Contents

Preface

This book is a culmination of my many years of practice in this field. I attribute the success of this book to my support group. I would like to thank my parents who have showered me with unconditional love and support and my peers and professors for their constant guidance.

The discipline which is concerned with the cultivation of ornamental as well as flowering plants for the floral industry is termed as floriculture. There are numerous types of crops which are dealt with under this field such as bedding plants, pot plants, houseplants and cut flowers. Some of the major flowering plants are orchids, poinsettias, florist chrysanthemums and florist azaleas. Numerous aspects of floriculture are applied in the farming of flowers such as spacing, training and pruning plants for optimal flower harvest. The diverse post-harvest treatments such as preservation, chemical treatment and packaging also fall within this field. This book covers in detail some existent theories and innovative practices revolving around floriculture. It presents this complex subject in the most comprehensible and easy to understand language. This book is appropriate for students seeking detailed information in this area as well as for experts.

The details of chapters are provided below for a progressive learning:

Chapter – Introduction

The discipline of horticulture which deals with the cultivation of flowering and ornamental plants for gardens and floristry is referred to as floriculture. It is also involved in the development of new varieties of plants through techniques such as plant breeding. This chapter has been carefully written to provide an easy understanding of the various aspects of floriculture as well as the structure of flowering plants.

Chapter – Flower: Cultivation and Uses

The reproductive portion of a flowering plant is called a flower. It generally facilitates the union of sperm with eggs. Some of the common types of flowers are lilium, jasmine, orchid, rose and hibiscus. The topics elaborated in this chapter will help in gaining a better perspective about the cultivation and uses of these flowers.

Chapter – Flower Structure, Morphology and Development

The structure of the flower consists of a non-reproductive part, known as a perianth, and the male and female reproductive parts. This chapter closely examines the different parts which constitute the structure of flowers as well as the morphology and development of flowers to provide an extensive understanding of the subject.

Chapter – Ornamental Plants and their Cultivars

The plants which are grown for display purposes instead of functional purposes are known as ornamental plants. They include various kinds of cultivars such as rose cultivars, orchid cultivars, banksia cultivars, grevillea cultivars and callistemon cultivars. This chapter has been carefully written to provide an easy understanding of ornamental plants and these cultivars.

Chapter – Ornamental Plant Pathogens and Diseases

Various pathogens such as cymbidium mosaic virus, orchid fleck virus, diplocarpon rosae and podosphaera pannosa can cause diseases in different ornamental plants. The topics elaborated in this chapter will help in gaining a better perspective about these pathogens and diseases which can afflict various ornamental plants such as roses and orchids.

Violet Dalton

1
Introduction

The discipline of horticulture which deals with the cultivation of flowering and ornamental plants for gardens and floristry is referred to as floriculture. It is also involved in the development of new varieties of plants through techniques such as plant breeding. This chapter has been carefully written to provide an easy understanding of the various aspects of floriculture as well as the structure of flowering plants.

Floriculture

Floriculture, also known as flower farming is a branch of horticulture that deals with cultivating ornamental and flowering plants. The flowers and plants cultivated are meant for sale. These can be used in the cosmetic industry, the perfume industry and even the pharmaceutical industry.

Floriculture not only includes the cultivation of plants but also their marketing. Flowers are marketed to local as well as distant markets. Cut flowers are also exported long with its products like scents, medicines and oils. The commercialization of flower cultivation has been a result of changing lifestyle of people.

Various forms of floriculture plants include bedding plants, foliage plants, cut flowers, flowering plants and cut cultivated greens. Flowering plants are used indoor and are sold in pots. Foliage plants are also used indoor and are sold in pots or hanging baskets. Cut flowers are sold in bouquets and bunches.

Importace of Floriculture

Flowers are considered a symbol of love, grace and elegance. We use flowers on religious occasions too. Flowers are given as birthday gifts, wedding gifts, at funerals and also when one goes to meet a sick person. Many Hindu ladies use flowers to style their hair in the form of gajras and veni. Apart from beautification and decoration, flowers have industrial importance too. Flowers like rose, jasmine give essential oils which are used in making perfumes and scents.

Flowering Plant

The flowering plants (also called angiosperms) are the dominant and most familiar group of land plants. The flowering plants and the gymnosperms comprise the two groups in the seed plants. The flowers of flowering plants are the most remarkable feature that distinguishes them from other seed plants.

Flowers initiated the differences between gymnosperms and angiosperms by broadening the scope of evolutionary relationships and niches open to flowering plants, allowing them to eventually dominate terrestrial ecosystems. The number of species of flowering plants is estimated to be in the range of 250, 000 to 400, 000.

Structure of Flowering Plants

Tap Roots

Taproots develop from the initial root that emerges from the seed. This was called the radicle. This is also called the primary root. These roots are present in most dicots.

Taproot of a carrot.

A lateral root, also called a secondary root is a side branch of the main root. The tips of these roots are covered with tiny root hairs.

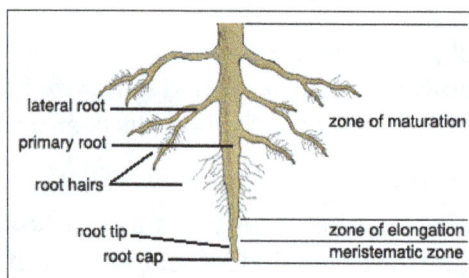

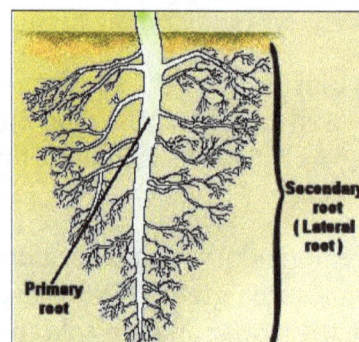

Root hairs.

Fibrous Roots

Most monocots have a fibrous root system consisting of an extensive mass of similarly sized roots. In these plants, the radicle is short lived and is replaced by a mass of equal sized roots. These roots are most common in monocots.

Fibrous roots of grass

Adventitious Roots

These are roots that do not grow from the radicle. They grow from various areas of the plant.

Functions of Roots

- Roots anchor the plant in the soil.

- Roots absorb water and mineral salts from the soil through their root hairs.

- Roots may store food. (carrots, turnips, radishes)

- Roots form a passageway for water and dissolved substances from the root into the stem and also for foods from the stem down into the root.

The Zones of a Root

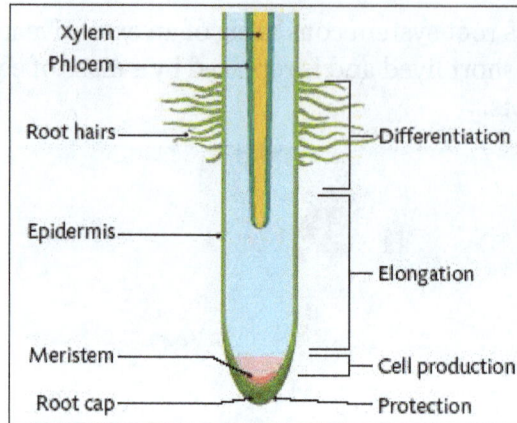

Zone of protection also known as the Root Cap: Envelope that protects the root as it pushes through the soil.

Meristem: Tissue at the tip of the root composed of rapidly multiplying cells. Meristem tissue is found in many areas of the plant. These are all tissues that have rapidly dividing cells for cell growth.

Elongation zone: Set of cells that determine the growth of the root. This area is where plant growth regulators, such as auxins, stimulate the cells from the meristem to grow larger.

Zone of Differentiation: This is the area where the elongated cells develop into different types of tissues. These types are:

- Dermal tissue: Such as epidermis that protects the plant.

- Ground tissue: Tissue found between the dermal and vascular tissue. Serves as structural strength.

- Vascular tissue: Tissue that transports material. Xylem transports water and phloem transports food.

Root tip:

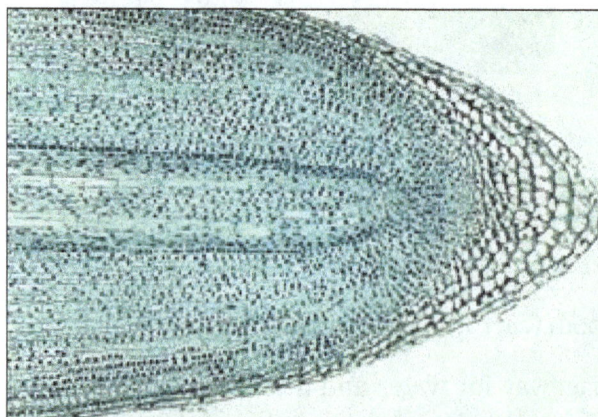

Stems

The stem is the main part of the shoot:

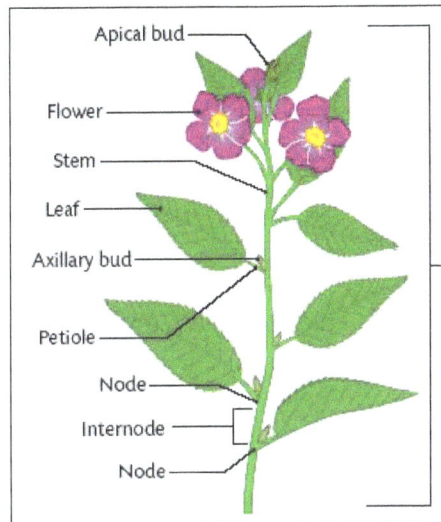

- Branch: A side stem that developed from an axillary bud.

- Petiole: Leaf stalk not always present.

- Node: The point of origin of a leaf on a stem.

- Internode: The section of stem between two successive internodes.

- Axil: The angle between the upper side of a leaf and its stem.

- Axillary Bud: Will develop into a side branch or flower.

- Apical or Terminal Bud: Increase in length of the stem forming new leaves and axillary buds.

- Lenticels: Loose open cork tissue for gas exchange for efficient aerobic respiration.

Function of Stems

- Formation of buds, leaves and flowers.

- Supports leaves in good light conditions to maximise photosynthesis.

- Vegetative reproduction e. g. stem tuber of potato.

- Food storage e. g. stem tuber of potato.

Leaves

The main part of the leaf is called the blade. The blade is attached to the stem at the node. The attachment is made? with the petiole of the leaf. The petiole becomes the midrib of the leaf. The petiole, midrib, and veins contain the xylem and phloem that carry food and water.

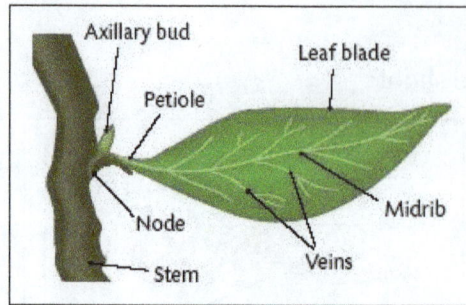

Venation

The pattern of veins on a leaf is called venation. There are 2 common types:

- Parallel Venation: The veins run along side each other. This venation is typical in monocots.

- Net or Reticulate Venation: The veins form a branching network. This venation is typical in dicots.

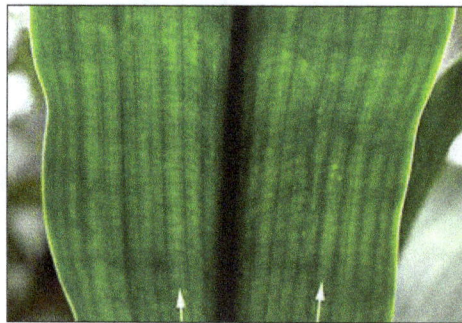

A comparison of both venations:

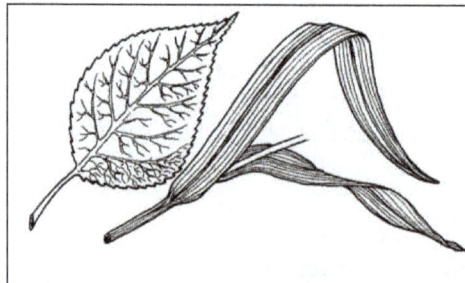

Functions of Leaves

Photosynthesis: The making of food in a plant. Through this process they obtain the food they need to live.

Gas exchange: They take in carbon dioxide and release oxygen.

Transpiration: They lose water through their leaves. As a result, fresh water enters the leaves through the midrib and veins.

Food Storage

Tissues in Flowering Plants

Plant tissues fall into three fundamental categories (among a few others): dermal tissue, ground tissue and vascular tissue.

Dermal Tissue

Dermal tissues generally occupy the "skin" layer of all plant organs. It is normally called epidermis. Its main function is protection of the plant. The epidermis of leaves is coated with a waxy cuticle to stop water loss.

Ground Tissue

This tissue occupies the space between the dermal tissues and the vascular tissues. These cells are much more than just filler, though. In roots the ground tissue may store sugars or starches to fuel the spring sap flow. In leaves, the ground tissue is that layer doing photosynthesis, the mesophyll. In many species ground tissues produce intracellular crystals that paralyse potential herbivores.

Vascular Tissues

The vascular tissues of higher plants (Kingdom Plantae) are divided into two sections: xylem and phloem. These two are found in the vascular bundles of plants.

Xylem- Transports water and minerals throughout a plant. It also provides support for the plant. Xylem consists of two types of cells.

- Vessels
- Tracheids

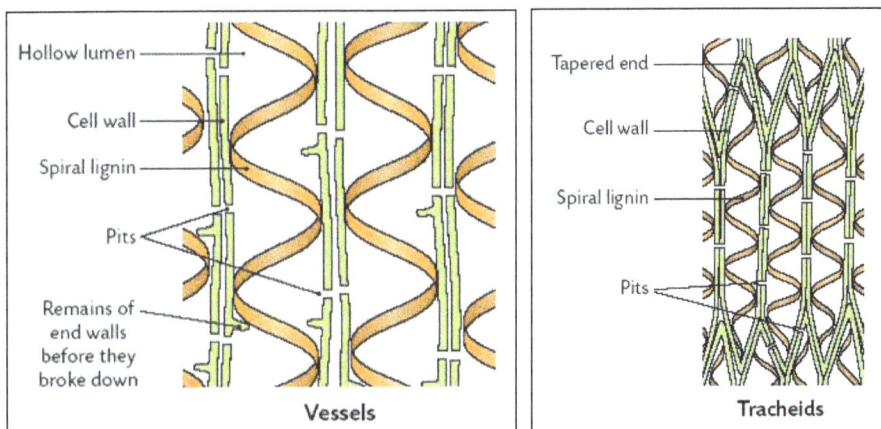

Both tracheids and vessels are tubes that allow water and dissolved minerals to pass throughout the plant. They also give support for the plant. They gain their strength because of lignin in their cell walls. The water and minerals go in and out of the tubes through the pits on the sides of the tubes. The vessels form continuous tubes while the tracheids are tapered at one end to allow one cell to overlap around another at their ends.

Tracheids are more primitive than vessels and are only found in coniferous trees such as pines.

Tracheid:

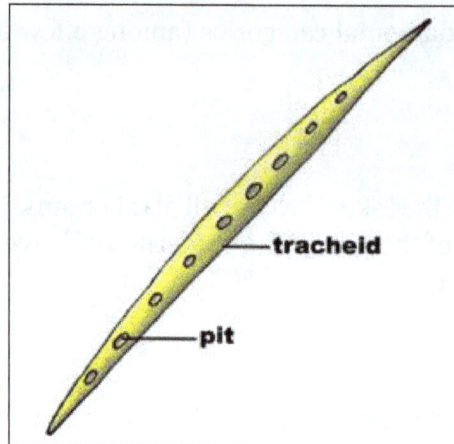

Vessel:

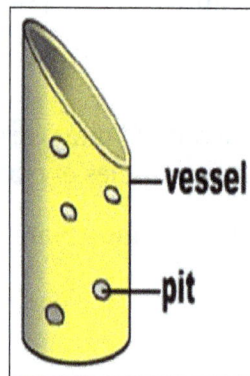

Xylem

- Tracheids and vessel members specialise in efficient water transport.
- Long, narrow, dead cells with walls thickened and strengthened with lignin.
- A series of vessel members forms a long continuous open tube called a xylem vessel.
- Pits in the thickened walls allow easy water transfer to neighbouring cells.
- Tracheids and vessel members also give great mechanical support to the plant.

Phloem

Phloem carries food that is made in the leaves of the plant. It transports the food to the rest of the plant. Phloem is composed of sieve tubes and companion cells.

Sieve tubes are long, tubular structures. They when individual sieve tube cells called sieve tube elements, join end-to end. The end walls develop pores that enable materials to pass through. These end walls are called sieve plates. The walls are made of cellulose, not lignin. Therefore, phloem is not as strong as xylem. These tubes have lost their nuclei.

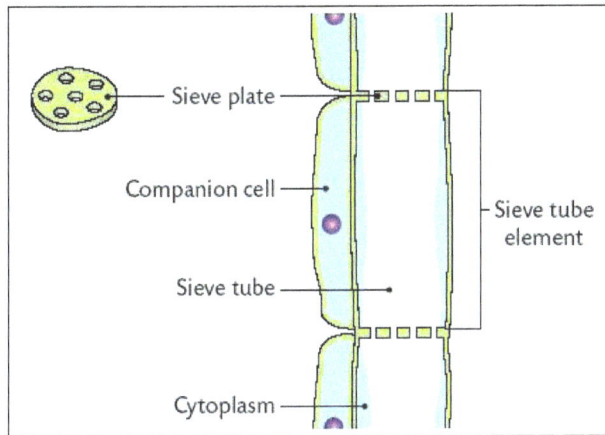

Each sieve tube element has a companion cell. These cells have a living cytoplasm and nucleus. Phloem, therefore, is a living tissue because of the living companion cell associated with each sieve tube element. Companion cells control the activities of the sieve tube element it is associated with.

- Specialises in efficient transport of food.
- Living cells but do not have a nucleus.
- Long, narrow, thin walled living cells.
- End walls are heavily perforated called a sieve plate.
- A series of sieve elements is called a sieve tube.

Companion Cells:

- Assist the sieve element in food transport.
- Live narrow cells with a prominent nucleus.
- Its nucleus also controls the sieve element.
- Dense cytoplasm particularly rich in mitochondria.

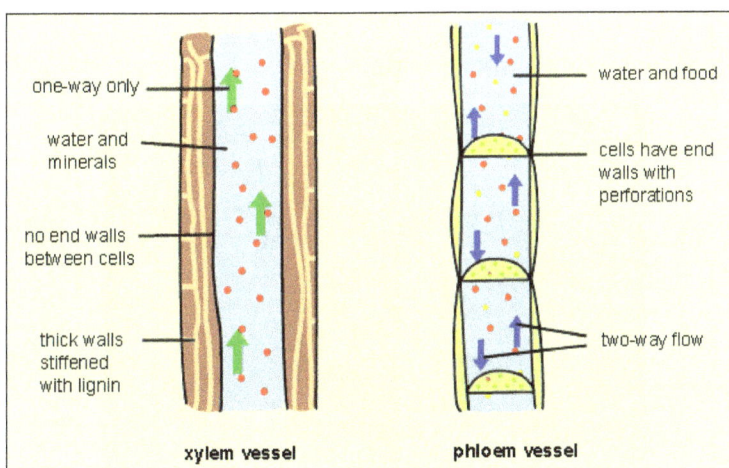

Location of Plant Tissues in Roots, Stems and Leaves

The following topic depicts the typical view of the dermal tissue, ground tissue, and vascular tissues (xylem and phloem in vascular bundles) in roots, stems, and leaves.

TS of Roots

This diagram is of a typical dicot root:

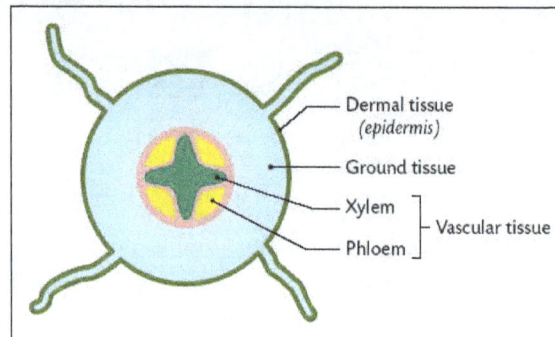

This is a diagram of a typical monocot root:

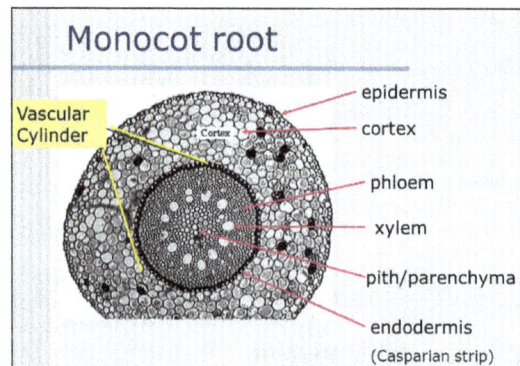

Dicot and monocots have different arrangements of root tissues. Notice that monocot roots have their xylem and phloem in a circular series around the root while dicots have them in one central location in the centre of the root.

This is a photo of a dicot root using a microscope:

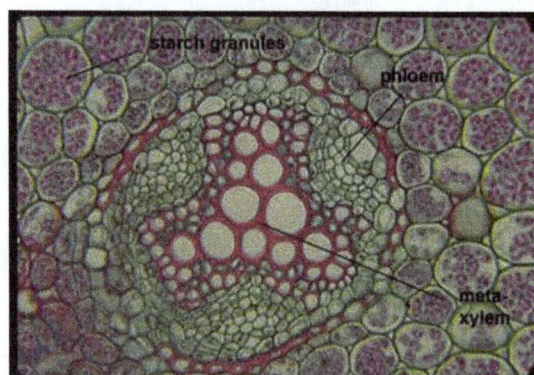

This is a photo of a monocot root using a microscope:

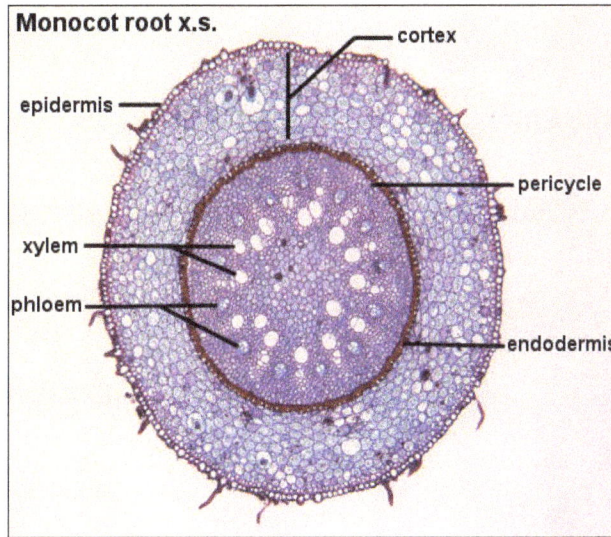

TS of Stems

This is a diagram of a typical dicot stem:

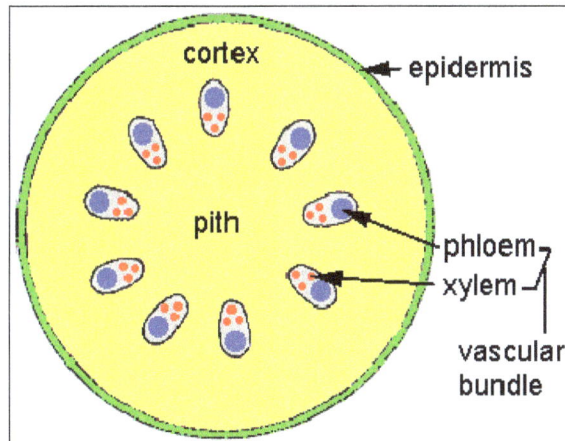

This is a diagram of a typical monocot stem:

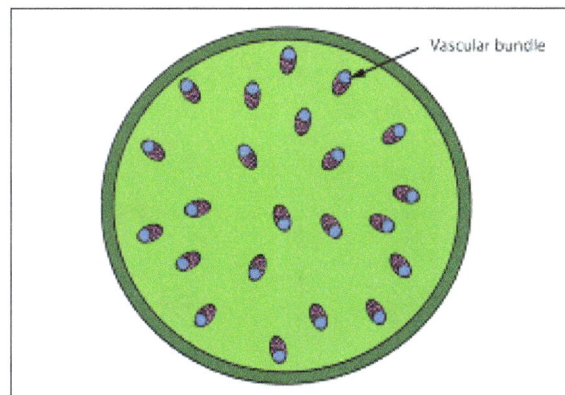

This is a photo of a dicot stem using a microscope:

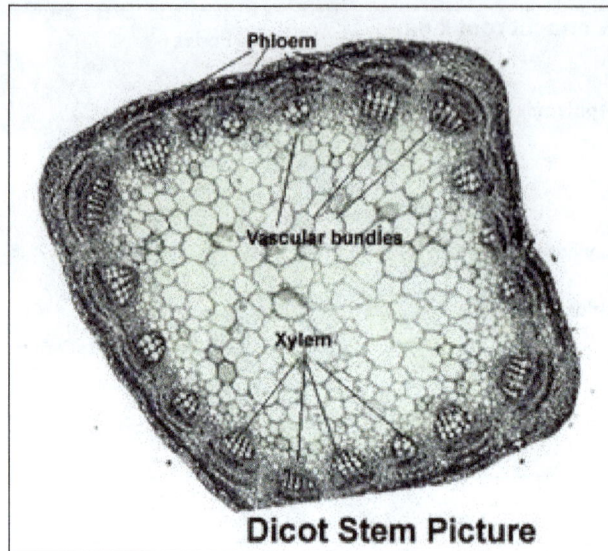

Dicot Stem Picture

This is a photo of a monocot stem using a microscope:

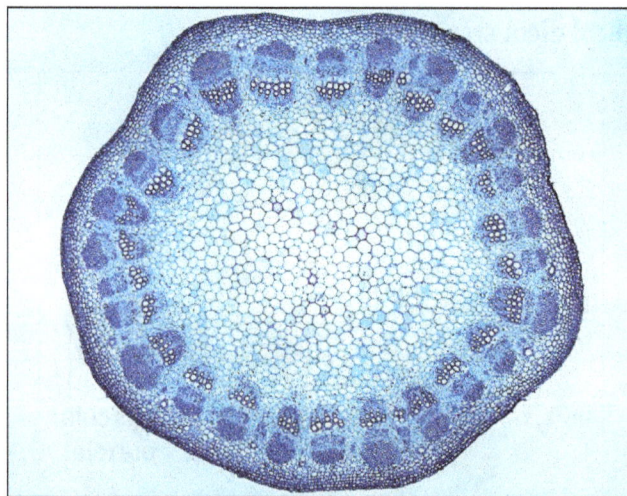

Stems-Long Section:

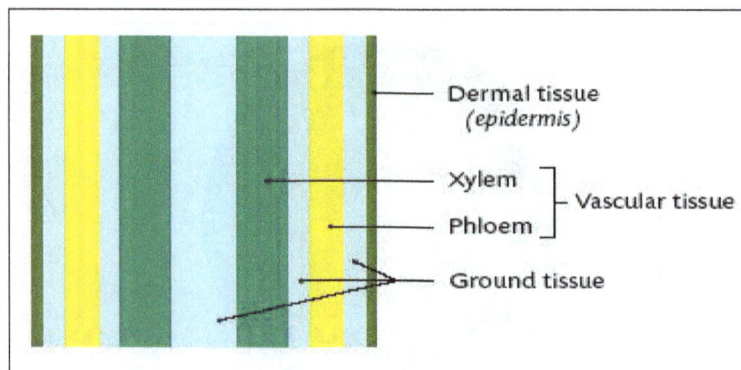

Leaf Cross Section:

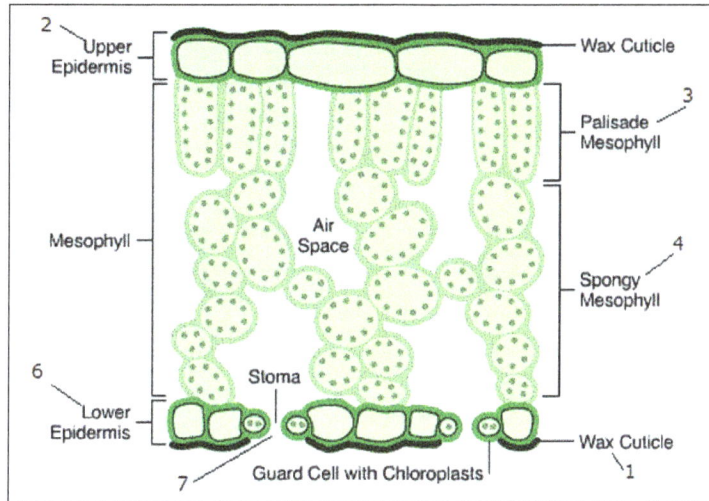

- Dermal Tissue: Outer single cell layer protective tissue.

- Cuticle: Layer of waterproof wax on the outer surface of the dermal tissue.

- Ground Tissue: Usually two layers, closely packed upper layer and loose lower layer photosynthetic tissue.

- Air Spaces: Rapid diffusion of carbon dioxide to the cells for photosynthesis.

- Guard Cells: Control the closing and opening of the stomatal pore.

- Stomata: Rapid entry of carbon dioxide into the leaf from the air.

Monocots and Dicots

Monocots and dicots get their names because of the number of cotyledons present in their seeds. Cotyledons are leaves in the seed that provide food for the seed embryo before it is able to develop its own leaves and make its own food (after germination).

Monocots have one seed leaf or cotyledon while dicots have 2 seed leaves or cotyledons.

Bean seed is a dicot and a corn seed is a monocot:

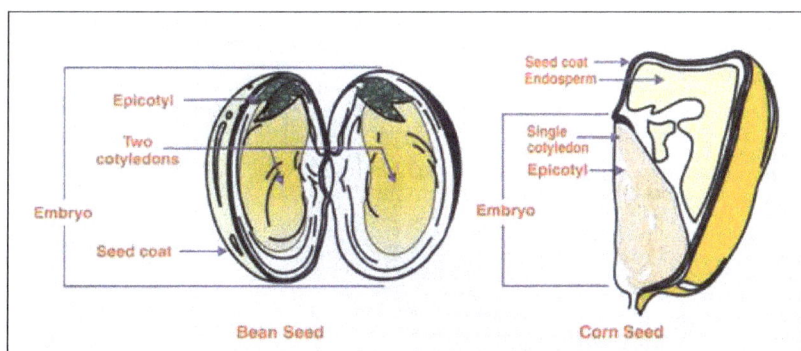

There are quite a few differences between a monocot and a dicot plant. This chart displays their differences:

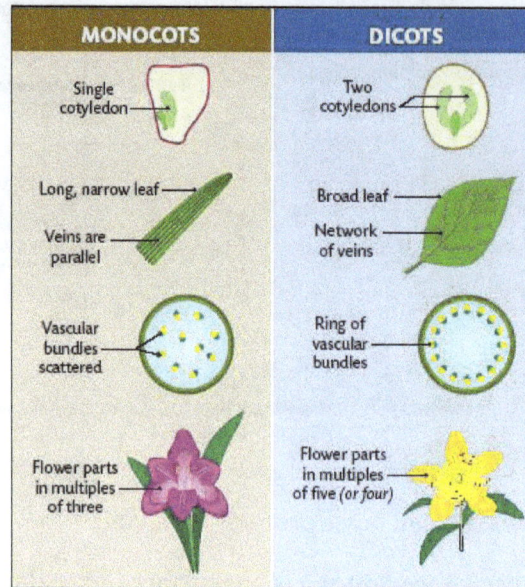

Mandatory Activity:

To prepare and examine a transverse section (TS) of a dicot stem:

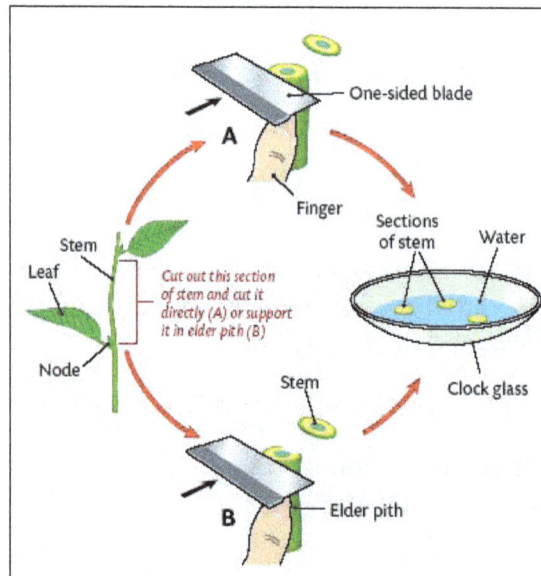

2

Flower: Cultivation and Uses

The reproductive portion of a flowering plant is called a flower. It generally facilitates the union of sperm with eggs. Some of the common types of flowers are lilium, jasmine, orchid, rose and hibiscus. The topics elaborated in this chapter will help in gaining a better perspective about the cultivation and uses of these flowers.

Flower

Flower is the reproductive portion of any plant in the division Magnoliophyta (Angiospermae), a group commonly called flowering plants or angiosperms. As popularly used, the term "flower" especially applies when part or all of the reproductive structure is distinctive in colour and form.

In their range of colour, size, form, and anatomical arrangement, flowers present a seemingly endless variety of combinations. They range in size from minute blossoms to giant blooms. In some plants, such as poppy, magnolia, tulip, and petunia, each flower is relatively large and showy and is produced singly, while in other plants, such as aster, snapdragon, calla lily, and lilac, the individual flowers may be very small and are borne in a distinctive cluster known as an inflorescence. Regardless of their variety, all flowers have a uniform function, the reproduction of the species through the production of seed. The flower is the characteristic structure of the evolutionarily highest group of plants, the angiosperms.

Petunia: Pink variegated flowers of a common
garden petunia (*Petunia ×atkinsiana*).

Common lilac (*Syringa vulgaris*).

Basically, each flower consists of a floral axis upon which are borne the essential organs of reproduction (stamens and pistils) and usually accessory organs (sepals and petals); the latter may serve to both attract pollinating insects and protect the essential organs. The floral axis is a greatly modified stem; unlike vegetative stems, which bear leaves, it is usually contracted, so that the parts of the flower are crowded together on the stem tip, the receptacle. The flower parts are usually arrayed in whorls (or cycles) but may also be disposed spirally, especially if the axis is elongate. There are commonly four distinct whorls of flower parts: (1) an outer calyx consisting of sepals; within it lies (2) the corolla, consisting of petals; (3) the androecium, or group of stamens; and in the centre is (4) the gynoecium, consisting of the pistils.

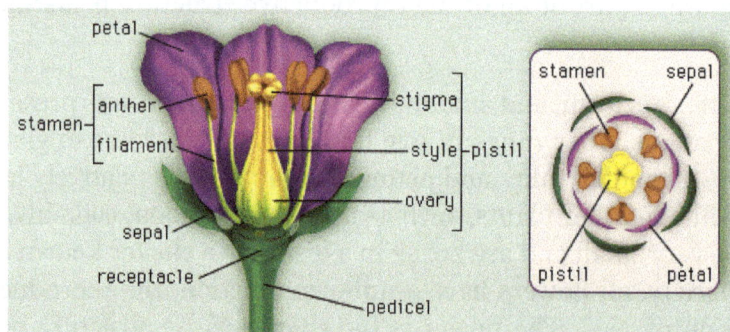

(Left) Generalized flower with parts; (right) diagram showing arrangement of floral parts in cross section at the flower's base

The sepals and petals together make up the perianth, or floral envelope. The sepals are usually greenish and often resemble reduced leaves, while the petals are usually colourful and showy. Sepals and petals that are indistinguishable, as in lilies and tulips, are sometimes referred to as tepals. The androecium, or male parts of the flower, comprise the stamens, each of which consists of a supporting filament and an anther, in which pollen is produced. The gynoecium, or female parts of the flower, comprise the pistils, each of which consists of an ovary, with an upright extension, the style, on the top of which rests the stigma, the pollen-receptive surface. The ovary encloses the ovules, or potential seeds. A pistil may be simple, made up of a single carpel, or ovule-bearing modified leaf; or compound, formed from several carpels joined together.

A flower having sepals, petals, stamens, and pistils is complete; lacking one or more of such structures, it is said to be incomplete. Stamens and pistils are not present together in all flowers. When both are present the flower is said to be perfect, or bisexual, regardless of a lack of any other part that renders it incomplete. A flower that lacks stamens is pistillate, or female, while one that lacks pistils is said to be staminate, or male. When the same plant bears unisexual flowers of both sexes, it is said to be monoecious (e. g., tuberous begonia, hazel, oak, corn); when the male and female flowers are on different plants, the plant is dioecious (e. g., date, holly, cottonwood, willow); when there are male, female, and bisexual flowers on the same plant, the plant is termed polygamous.

A perfect flower with floral structures in multiples of three, Tulipa (tulip) has a three-lobed stigma, six stamens, and six distinct perianth parts.

A flower may be radially symmetrical, as in roses and petunias, in which case it is termed regular or actinomorphic. A bilaterally symmetrical flower, as in orchids and snapdragons, is irregular or zygomorphic.

The radiate head of the treasure flower (Gazania rigens), a daisylike inflorescence composed of disk flowers in the centre surrounded by marginal ray flowers.

Neither the calyx nor the corolla is necessary for reproduction. The stamens and pistils, on the other hand, are directly involved with the production of seed. The stamen bears microsporangia (spore cases) in which are developed numerous microspores (potential pollen grains); the pistil bears ovules, each enclosing an egg cell. When a microspore germinates, it is known as a pollen grain. When the pollen sacs in a stamen's anther are ripe, the anther releases them and the pollen is shed. Fertilization can occur only if the pollen grains are transferred from the anther to the stigma of a pistil, a process known as pollination. This is of two chief kinds: (1) self-pollination, the pollination of a stigma by pollen from the same flower or another flower on the same plant;

and (2) cross-pollination, the transfer of pollen from the anther of a flower of one plant to the stigma of the flower of another plant of the same species. Self-pollination occurs in many species, but in the others, perhaps the majority, it is prevented by such adaptations as the structure of the flower, self-incompatibility, and the maturation of stamens and pistils of the same flower or plant at different times. Cross-pollination may be brought about by a number of agents, chiefly insects and wind. Wind-pollinated flowers generally can be recognized by their lack of colour, odour, or nectar, while insect-pollinated flowers are conspicuous by virtue of their structure, colour, or the production of scent or nectar.

Bilateral symmetry of the orchid (*Vanda*).

After a pollen grain has reached the stigma, it germinates, and a pollen tube protrudes from it. This tube, containing two male gametes (sperms), extends into the ovary and reaches the ovule, discharging its gametes so that they fertilize the egg cell, which becomes an embryo. (Normally many pollen grains fall on a stigma; they all may germinate, but only one pollen tube enters any one ovule.) Following fertilization, the embryo is on its way to becoming a seed, and at this time the ovary itself enlarges to form the fruit.

Flowers have been symbols of beauty in most civilizations of the world, and flower giving is still among the most popular of social amenities. As gifts, flowers serve as expressions of affection for spouses, other family members, and friends; as decorations at weddings and other ceremonies; as tokens of respect for the deceased; as cheering gifts to the bedridden; or as expressions of thanks to hostesses and other social contacts. Most flowers bought by the public are grown in commercial greenhouses and then sold through wholesalers to retail florists.

Lilium

Lilium (members of which are true lilies) is a genus of herbaceous flowering plants growing from bulbs, all with large prominent flowers. Lilies are a group of flowering plants which are important in culture and literature in much of the world. Most species are native to the temperate northern hemisphere, though their range extends into the northern subtropics. Many other plants have "lily" in their common name but are not related to true lilies.

Lilium longiflorum flower – 1. Stigma,
2. Style, 3. Stamens, 4. Filament, 5. Tepal.

Lilies are tall perennials ranging in height from 2–6 ft (60–180 cm). They form naked or tunicless scaly underground bulbs which are their organs of perennation. In some North American species the base of the bulb develops into rhizomes, on which numerous small bulbs are found. Some species develop stolons. Most bulbs are buried deep in the ground, but a few species form bulbs near the soil surface. Many species form stem-roots. With these, the bulb grows naturally at some depth in the soil, and each year the new stem puts out adventitious roots above the bulb as it emerges from the soil. These roots are in addition to the basal roots that develop at the base of the bulb.

Lily, petal.

The flowers are large, often fragrant, and come in a wide range of colors including whites, yellows, oranges, pinks, reds and purples. Markings include spots and brush strokes. The plants are late spring- or summer-flowering. Flowers are borne in racemes or umbels at the tip of the stem, with six tepals spreading or reflexed, to give flowers varying from funnel shape to a "Turk's cap". The tepals are free from each other, and bear a nectary at the base of each flower. The ovary is 'superior', borne above the point of attachment of the anthers. The fruit is a three-celled capsule.

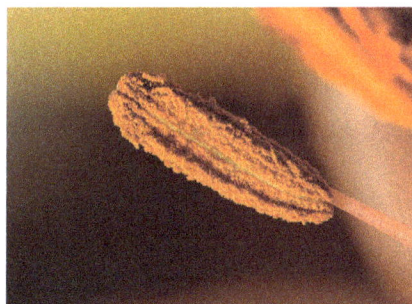

Stamen of lilium.

Seeds ripen in late summer. They exhibit varying and sometimes complex germination patterns, many adapted to cool temperate climates.

Naturally most cool temperate species are deciduous and dormant in winter in their native environment. But a few species which distribute in hot summer and mild winter area (Lilium candidum, Lilium catesbaei, Lilium longiflorum) lose leaves and remain relatively short dormant in Summer or Autumn, sprout from Autumn to winter, forming dwarf stem bearing a basal rosette of leaves until, after they have received sufficient chilling, the stem begins to elongate in warming weather.

Lilium candidum seeds.

The basic chromosome number is twelve (n=12).

Distribution and Habitat

The range of lilies in the Old World extends across much of Europe, across most of Asia to Japan, south to India, and east to Indochina and the Philippines. In the New World they extend from southern Canada through much of the United States. They are commonly adapted to either woodland habitats, often montane, or sometimes to grassland habitats. A few can survive in marshland and epiphytes are known in tropical southeast Asia. In general they prefer moderately acidic or lime-free soils.

Ecology

Lilies are used as food plants by the larvae of some Lepidoptera species including the Dun-bar.

Cultivation

Many species are widely grown in the garden in temperate and sub-tropical regions. They may also be grown as potted plants. Numerous ornamental hybrids have been developed. They can be used in herbaceous borders, woodland and shrub plantings, and as patio plants. Some lilies, especially Lilium longiflorum, form important cut flower crops. These may be forced for particular markets; for instance, Lilium longiflorum for the Easter trade, when it may be called the Easter lily.

Lilies are usually planted as bulbs in the dormant season. They are best planted in a south-facing (northern hemisphere), slightly sloping aspect, in sun or part shade, at a depth 2½ times the height of the bulb (except Lilium candidum which should be planted at the surface). Most prefer a porous, loamy soil, and good drainage is essential. Most species bloom in July or August (northern hemisphere). The flowering periods of certain lily species begin in late spring, while others bloom in late summer or early autumn. They have contractile roots which pull the plant down to the correct depth, therefore it is better to plant them too shallowly than too deep. A soil pH of around 6.5 is generally safe. The soil should be well-drained, and plants must be kept watered during the growing season. Some plants have strong wiry stems, but those with heavy flower heads may need staking.

Awards

'Golden Splendor'.

The following lily species and cultivars currently hold the Royal Horticultural Society's Award of Garden Merit:

- African Queen Group – (VI-/a) 2002 H6
- 'Casa Blanca' – (VIIb/b-c) 1993 H6
- 'Fata Morgana' – (Ia/b) 2002 H6
- 'Garden Party' – (VIIb/b) 2002 H6
- Golden Splendor Group – (VIb-c/a)
- Lilium henryi – (IXc/d) 1993 H6
- Lilium mackliniae – (IXc/a) 2012 H5
- Lilium martagon – Turk's cap lily (IXc/d)
- Lilium pardalinum – leopard lily (IXc/d)
- Pink Perfection Group – (VIb/a)
- Lilium regale – regal lily, king's lily (IXb/a)

Classification of Garden Forms

Numerous forms, mostly hybrids, are grown for the garden. They vary according to the species and interspecific hybrids that they derived from, and are classified in the following broad groups:

Asiatic Hybrids (Division I)

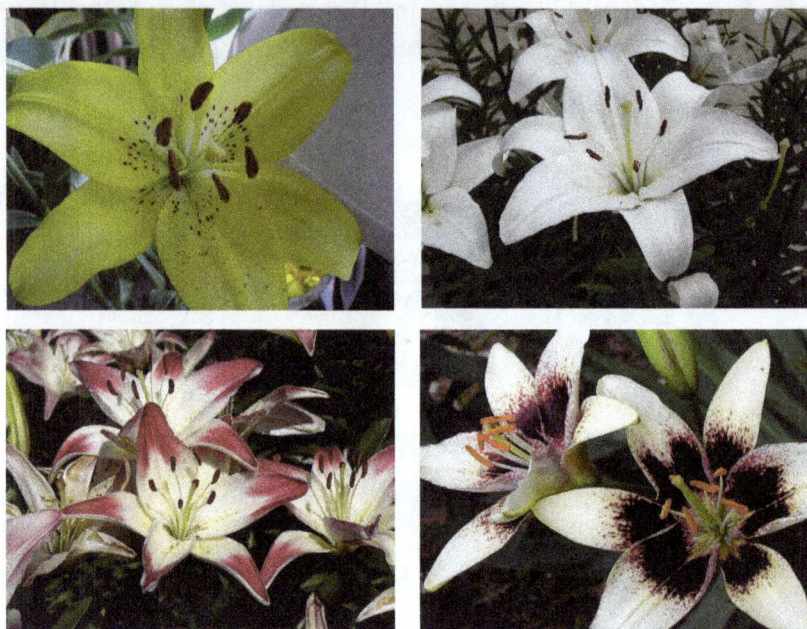

These are derived from hybrids between species in Lilium section Sinomartagon.

They are derived from central and East Asian species and interspecific hybrids, including Lilium amabile, Lilium bulbiferum, Lilium callosum, Lilium cernuum, Lilium concolor, Lilium dauricum, Lilium davidii, Lilium × hollandicum, Lilium lancifolium (syn. Lilium tigrinum), Lilium lankongense, Lilium leichtlinii, Lilium × maculatum, Lilium pumilum, Lilium × scottiae, Lilium wardii and Lilium wilsonii.

These are plants with medium-sized, upright or outward facing flowers, mostly unscented. There are various cultivars such as Lilium 'Cappuccino', Lilium 'Dimension', Lilium 'Little Kiss' and Lilium 'Navona'.

- Dwarf (Patio, Border) varieties are much shorter, c.36–61 cm in height and were designed for containers. They often bear the cultivar name 'Tiny', such as the 'Lily Looks' series, e. g. 'Tiny Padhye', 'Tiny Dessert'.

Martagon Hybrids (Division II)

These are based on Lilium dalhansonii, Lilium hansonii, Lilium martagon, Lilium medeoloides, and Lilium tsingtauense.

The flowers are nodding, Turk's cap style (with the petals strongly recurved).

Candidum (Euro-caucasian) Hybrids (Division III)

This includes mostly European species: Lilium candidum, Lilium chalcedonicum, Lilium kesselringianum, Lilium monadelphum, Lilium pomponium, Lilium pyrenaicum and Lilium × testaceum.

American Hybrids (Division IV)

These are mostly taller growing forms, originally derived from Lilium bolanderi, Lilium × burbankii, Lilium canadense, Lilium columbianum, Lilium grayi, Lilium humboldtii, Lilium kelleyanum, Lilium kelloggii, Lilium maritimum, Lilium michauxii, Lilium michiganense, Lilium occidentale, Lilium × pardaboldtii, Lilium pardalinum, Lilium parryi, Lilium parvum, Lilium philadelphicum, Lilium pitkinense, Lilium superbum, Lilium ollmeri, Lilium washingtonianum, and Lilium wigginsii.

Many are clump-forming perennials with rhizomatous rootstocks.

Longiflorum Hybrids (Division V)

These are cultivated forms of this species and its subspecies.

They are most important as plants for cut flowers, and are less often grown in the garden than other hybrids.

Trumpet Lilies (Division VI), Including Aurelian Hybrids (with L. Henryi)

This group includes hybrids of many Asiatic species and their interspecific hybrids, including Lilium × aurelianense, Lilium brownii, Lilium × centigale, Lilium henryi, Lilium × imperiale, Lilium × kewense, Lilium leucanthum, Lilium regale, Lilium rosthornii, Lilium sargentiae, Lilium sulphureum and Lilium × sulphurgale.

The flowers are trumpet shaped, facing outward or somewhat downward, and tend to be strongly fragrant, often especially night-fragrant.

Oriental Hybrids (Division VII)

These are based on hybrids within Lilium section Archelirion, specifically Lilium auratum and Lilium speciosum, together with crossbreeds from several species native to Japan, including Lilium nobilissimum, Lilium rubellum, Lilium alexandrae, and Lilium japonicum.

They are fragrant, and the flowers tend to be outward facing. Plants tend to be tall, and the flowers may be quite large. The whole group are sometimes referred to as "stargazers" because many of them appear to look upwards.

Other Hybrids (Division VIII)

Includes all other garden hybrids.

Species (Division IX)

All natural species and naturally occurring forms are included in this group.

The flowers can be classified by flower aspect and form:

- Flower aspect:
  - a up-facing
  - b out-facing
  - c down-facing

- Flower form:
  - a trumpet-shaped
  - b bowl-shaped
  - c flat (or with tepal tips recurved)
  - d tepals strongly recurved (with the Turk's cap form as the ultimate state)

Many newer commercial varieties are developed by using new technologies such as ovary culture and embryo rescue.

Pests and Diseases

Scarlet lily beetles, Oxfordshire, UK

Aphids may infest plants. Leatherjackets feed on the roots. Larvae of the Scarlet lily beetle can cause serious damage to the stems and leaves. The scarlet beetle lays its eggs and completes its life cycle only on true lilies (Lilium) and fritillaries (Fritillaria). Oriental, rubrum, tiger and trumpet lilies as well as Oriental trumpets (orienpets) and Turk's cap lilies and native North American Lilium species are all vulnerable, but the beetle prefers some types over others. The beetle could also be having an effect on native Canadian species and some rare and endangered species found in northeastern North America. Daylilies (Hemerocallis, not true lilies) are excluded from this category. Plants can suffer from damage caused by mice, deer and squirrels. Slugs, snails and millipedes attack seedlings, leaves and flowers. Brown spots on damp leaves may signal botrytis (also known as lily disease). Various fungal and viral diseases can cause mottling of leaves and stunting of growth.

Propagation and Growth

Lilies can be propagated in several ways;

- By division of the bulbs.
- By growing-on bulbils which are adventitious bulbs formed on the stem.
- By scaling, for which whole scales are detached from the bulb and planted to form a new bulb.
- By seed; there are many seed germination patterns, which can be complex.

- By micropropagation techniques (which include tissue culture); commercial quantities of lilies are often propagated in vitro and then planted out to grow into plants large enough to sell.

Lily growing in Eastern Siberia.

According to a study done by Anna Pobudkiewicz and Jadwiga the use of flurprimidol foliar spray helps aid in the limitation of stem elongation in oriental lilies.

Jasmine

Jasmine (taxonomic name Jasminum) is a genus of shrubs and vines in the olive family (Oleaceae). It contains around 200 species native to tropical and warm temperate regions of Eurasia and Oceania. Jasmines are widely cultivated for the characteristic fragrance of their flowers. A number of unrelated plants contain the word "Jasmine" in their common names.

Jasmine can be either deciduous (leaves falling in autumn) or evergreen (green all year round), and can be erect, spreading, or climbing shrubs and vines. Their leaves are borne in opposing or alternating arrangement and can be of simple, trifoliate, or pinnate formation. The flowers are typically around 2.5 cm (0.98 in) in diameter. They are white or yellow in color, although in rare instances they can be slightly reddish. The flowers are borne in cymose clusters with a minimum of three flowers, though they can also be solitary on the ends of branchlets. Each flower has about four to nine petals, two locules, and one to four ovules. They have two stamens with very short filaments. The bracts are linear or ovate. The calyx is bell-shaped. They are usually very fragrant. The fruits of jasmines are berries that turn black when ripe. The basic chromosome number of the genus is 13, and most species are diploid (2n=26). However, natural polyploidy exists, particularly in Jasminum sambac (2n=39), Jasminum flexile (2n=52), Jasminum mesnyi (2n=39), and Jasminum angustifolium (2n=52).

Distribution and Habitat

Jasmines are native to tropical and subtropical regions of Eurasia, Australasia and Oceania, although only one of the 200 species is native to Europe. Their center of diversity is in South Asia and Southeast Asia.

A number of jasmine species have become naturalized in Mediterranean Europe. For example, the so-called Spanish jasmine (Jasminum grandiflorum) was originally from West Asia and Indian subcontinent, and is now naturalized in the Iberian peninsula.

Jasminum fluminense (which is sometimes known by the inaccurate name "Brazilian Jasmine") and Jasminum dichotomum (Gold Coast Jasmine) are invasive species in Hawaii and Florida. Jasminum polyanthum, also known as White Jasmine, is an invasive weed in Australia.

Cultivation and Uses

Widely cultivated for its flowers, jasmine is enjoyed in the garden, as a house plant, and as cut flowers. The flowers are worn by women in their hair in South and South East Asia.

Jasmine Tea

Green tea with jasmine flowers.

Jasmine tea is often consumed in China, where it is called jasmine-flower tea. Jasminum sambac flowers are also used to make jasmine tea, which often has a base of green tea or white tea, but sometimes an Oolong base is used. The flowers are put in machines that control temperature and humidity. It takes about four hours for the tea to absorb the fragrance and flavour of the jasmine blossoms. For the highest grades of jasmine tea, this process may be repeated up to seven times. As the tea absorbs moisture from the fresh Jasmine flowers, it must be refired to prevent spoilage. The used flowers may be removed from the final product, as the flowers contain no more aroma. Giant fans are used to blow away and remove the petals from the denser tea leaves.

In Okinawa, Japan, jasmine tea is known as sanpin cha.

Jasmonates

Jasmine gave name to the jasmonate plant hormones, as methyl jasmonate isolated from the oil of Jasminum grandiflorum led to the discovery of the molecular structure of jasmonates. Jasmonates occur ubiquitously across the plant kingdom, having key roles in responses to environmental cues, such as heat or cold stress, and participate in the signal transduction pathways of many plants.

Plantation

Jasmine plantation is usually done using the stem of an existing plant, or one having roots. In rare occasions, the flowers bear dark purple fruits with seeds. The seeds germinated when sowed and nurtured properly. The flowering shrubs are usually trimmed pre-summer, as fresh branches grow and bear flowers during the summer.

Orchid

The Orchidaceae are a diverse and widespread family of flowering plants, with blooms that are often colourful and fragrant, commonly known as the orchid family.

Along with the Asteraceae, they are one of the two largest families of flowering plants. The Orchidaceae have about 28, 000 currently accepted species, distributed in about 763 genera. The determination of which family is larger is still under debate, because verified data on the members of such enormous families are continually in flux. Regardless, the number of orchid species nearly equals the number of bony fishes and is more than twice the number of bird species, and about four times the number of mammal species.

The family encompasses about 6–11% of all seed plants. The largest genera are Bulbophyllum (2, 000 species), Epidendrum (1, 500 species), Dendrobium (1, 400 species) and Pleurothallis (1, 000 species). It also includes Vanilla–the genus of the vanilla plant, the type genus Orchis, and many commonly cultivated plants such as Phalaenopsis and Cattleya. Moreover, since the introduction of tropical species into cultivation in the 19th century, horticulturists have produced more than 100, 000 hybrids and cultivars.

High resolution image of orchid.

Orchids are easily distinguished from other plants, as they share some very evident, shared derived characteristics, or synapomorphies. Among these are: bilateral symmetry of the flower (zygomorphism), many resupinate flowers, a nearly always highly modified petal (labellum), fused stamens and carpels, and extremely small seeds.

Stem and Roots

Germinating seeds of the temperate orchid Anacamptis coriophora.
The protocorm is the first organ that will develop into true roots and leaves.

All orchids are perennial herbs that lack any permanent woody structure. They can grow according to two patterns:

- Monopodial: The stem grows from a single bud, leaves are added from the apex each year and the stem grows longer accordingly. The stem of orchids with a monopodial growth can reach several metres in length, as in Vanda and Vanilla.

- Sympodial: Sympodial orchids have a front (the newest growth) and a back (the oldest growth). The plant produces a series of adjacent shoots, which grow to a certain size, bloom and then stop growing and are replaced. Sympodial orchids grow laterally rather than vertically, following the surface of their support. The growth continues by development of new leads, with their own leaves and roots, sprouting from or next to those of the previous year, as in Cattleya. While a new lead is developing, the rhizome may start its growth again from a so-called 'eye', an undeveloped bud, thereby branching. Sympodial orchids may have visible pseudobulbs joined by a rhizome, which creeps along the top or just beneath the soil.

Anacamptis lactea showing the two tubers.

Terrestrial orchids may be rhizomatous or form corms or tubers. The root caps of terrestrial orchids are smooth and white. Some sympodial terrestrial orchids, such as *Orchis* and *Ophrys*, have two subterranean tuberous roots. One is used as a food reserve for wintry periods, and provides for the development of the other one, from which visible growth develops.

In warm and constantly humid climates, many terrestrial orchids do not need pseudobulbs.

Epiphytic orchids, those that grow upon a support, have modified aerial roots that can sometimes be a few meters long. In the older parts of the roots, a modified spongy epidermis, called a velamen, has the function of absorbing humidity. It is made of dead cells and can have a silvery-grey, white or brown appearance. In some orchids, the velamen includes spongy and fibrous bodies near the passage cells, called tilosomes.

The cells of the root epidermis grow at a right angle to the axis of the root to allow them to get a firm grasp on their support. Nutrients for epiphytic orchids mainly come from mineral dust, organic detritus, animal droppings and other substances collecting among on their supporting surfaces.

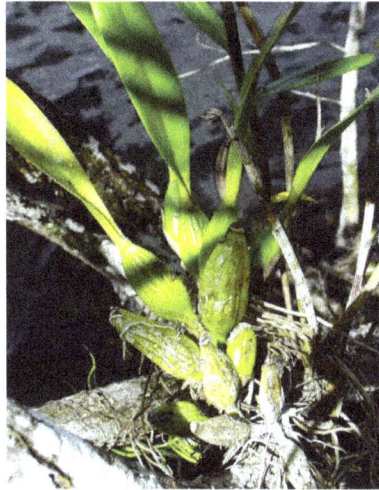

The pseudobulb of Prosthechea fragrans.

The base of the stem of sympodial epiphytes, or in some species essentially the entire stem, may be thickened to form a pseudobulb that contains nutrients and water for drier periods.

The pseudobulb has a smooth surface with lengthwise grooves, and can have different shapes, often conical or oblong. Its size is very variable; in some small species of Bulbophyllum, it is no longer than two millimeters, while in the largest orchid in the world, Grammatophyllum speciosum (giant orchid), it can reach three meters. Some Dendrobium species have long, canelike pseudobulbs with short, rounded leaves over the whole length; some other orchids have hidden or extremely small pseudobulbs, completely included inside the leaves.

With ageing, the pseudobulb sheds its leaves and becomes dormant. At this stage, it is often called a backbulb. Backbulbs still hold nutrition for the plant, but then a pseudobulb usually takes over, exploiting the last reserves accumulated in the backbulb, which eventually dies off, too. A pseudobulb typically lives for about five years. Orchids without noticeable pseudobulbs are also said to have growths, an individual component of a sympodial plant.

Leaves

Like most monocots, orchids generally have simple leaves with parallel veins, although some Vanilloideae have reticulate venation. Leaves may be ovate, lanceolate, or orbiculate, and very variable in size on the individual plant. Their characteristics are often diagnostic. They are normally alternate on the stem, often folded lengthwise along the centre ("plicate"), and have no stipules. Orchid leaves often have siliceous bodies called stegmata in the vascular bundle sheaths (not present in the Orchidoideae) and are fibrous.

The structure of the leaves corresponds to the specific habitat of the plant. Species that typically bask in sunlight, or grow on sites which can be occasionally very dry, have thick, leathery leaves and the laminae are covered by a waxy cuticle to retain their necessary water supply. Shade-loving species, on the other hand, have long, thin leaves.

The leaves of most orchids are perennial, that is, they live for several years, while others, especially those with plicate leaves as in Catasetum, shed them annually and develop new leaves together with new pseudobulbs.

The leaves of some orchids are considered ornamental. The leaves of the Macodes sanderiana, a semiterrestrial or rock-hugging ("lithophyte") orchid, show a sparkling silver and gold veining on a light green background. The cordate leaves of Psychopsis limminghei are light brownish-green with maroon-puce markings, created by flower pigments. The attractive mottle of the leaves of lady's slippers from tropical and subtropical Asia (Paphiopedilum), is caused by uneven distribution of chlorophyll. Also, Phalaenopsis schilleriana is a pastel pink orchid with leaves spotted dark green and light green. The jewel orchid (Ludisia discolor) is grown more for its colorful leaves than its white flowers.

Vanda cultivar.

Some orchids, such as Dendrophylax lindenii (ghost orchid), Aphyllorchis and Taeniophyllum depend on their green roots for photosynthesis and lack normally developed leaves, as do all of the heterotrophic species.

Orchids of the genus Corallorhiza (coralroot orchids) lack leaves altogether and instead wrap their roots around the roots of mature trees and use specialized fungi to harvest sugars.

Flowers

The Orchidaceae are well known for the many structural variations in their flowers. Some orchids have single flowers, but most have a racemose inflorescence, sometimes with a large number of flowers. The flowering stem can be basal, that is, produced from the base of the tuber, like in Cymbidium, apical, meaning it grows from the apex of the main stem, like in Cattleya, or axillary, from the leaf axil, as in Vanda.

As an apomorphy of the clade, orchid flowers are primitively zygomorphic (bilaterally symmetrical), although in some genera, such as Mormodes, Ludisia, and Macodes, this kind of symmetry may be difficult to notice.

Dactylorhiza sambucina.

The orchid flower, like most flowers of monocots, has two whorls of sterile elements. The outer whorl has three sepals and the inner whorl has three petals. The sepals are usually very similar to the petals (thus called tepals), but may be completely distinct.

The medial petal, called the labellum or lip, which is always modified and enlarged, is actually the upper medial petal; however, as the flower develops, the inferior ovary or the pedicel usually rotates 180°, so that the labellum arrives at the lower part of the flower, thus becoming suitable to form a platform for pollinators. This characteristic, called resupination, occurs primitively in the family and is considered apomorphic, a derived characteristic all Orchidaceae share. The torsion of the ovary is very evident from the longitudinal section shown. Some orchids have secondarily lost this resupination, e. g. Epidendrum secundum.

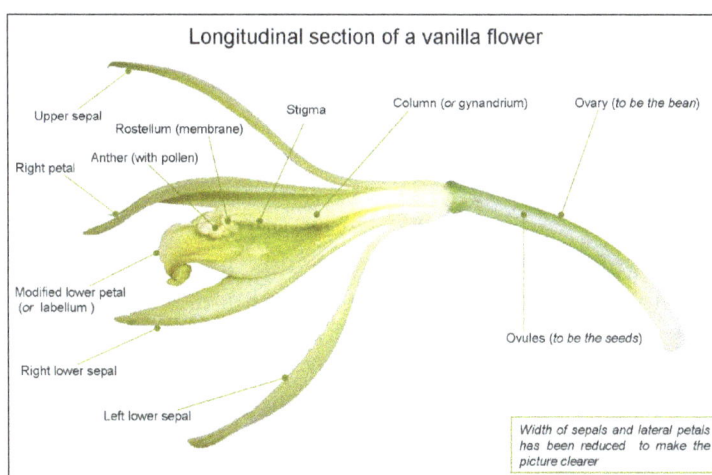

Longitudinal section of a flower of *Vanilla planifolia*.

The normal form of the sepals can be found in Cattleya, where they form a triangle. In Paphiopedilum (Venus slippers), the lower two sepals are fused into a synsepal, while the lip has taken the form of a slipper. In Masdevallia, all the sepals are fused.

Orchid flowers with abnormal numbers of petals or lips are called peloric. Peloria is a genetic trait, but its expression is environmentally influenced and may appear random.

Laeliocattleya cultivar shows the normal form of petals.

Orchid flowers primitively had three stamens, but this situation is now limited to the genus Neuwiedia. Apostasia and the Cypripedioideae have two stamens, the central one being sterile and reduced to a staminode. All of the other orchids, the clade called Monandria, retain only the central stamen, the others being reduced to staminodes. The filaments of the stamens are always adnate (fused) to the style to form cylindrical structure called the gynostemium or column. In the primitive Apostasioideae, this fusion is only partial; in the Vanilloideae, it is more deep; in Orchidoideae and Epidendroideae, it is total. The stigma is very asymmetrical, as all of its lobes are bent towards the centre of the flower and lie on the bottom of the column.

Pollen is released as single grains, like in most other plants, in the Apostasioideae, Cypripedioideae, and Vanilloideae. In the other subfamilies, which comprise the great majority of orchids, the anther carries two pollinia.

A pollinium is a waxy mass of pollen grains held together by the glue-like alkaloid viscin, containing both cellulosic strands and mucopolysaccharides. Each pollinium is connected to a filament which can take the form of a caudicle, as in Dactylorhiza or Habenaria, or a stipe, as in Vanda. Caudicles or stipes hold the pollinia to the viscidium, a sticky pad which sticks the pollinia to the body of pollinators.

At the upper edge of the stigma of single-anthered orchids, in front of the anther cap, is the rostellum, a slender extension involved in the complex pollination mechanism.

The ovary is always inferior (located behind the flower). It is three-carpelate and one or, more rarely, three-partitioned, with parietal placentation (axile in the Apostasioideae).

In 2011, Bulbophyllum nocturnum was discovered to flower nocturnally.

Pollination

The complex mechanisms that orchids have evolved to achieve cross-pollination were investigated by Charles Darwin and described in Fertilisation of Orchids (1862). Orchids have developed highly specialized pollination systems, thus the chances of being pollinated are often scarce, so orchid flowers usually remain receptive for very long periods, rendering unpollinated flowers long-lasting in cultivation. Most orchids deliver pollen in a single mass. Each time pollination succeeds, thousands of ovules can be fertilized.

Pollinators are often visually attracted by the shape and colours of the labellum. However, some *Bulbophyllum* species attract male fruit flies (*Bactrocera* spp.) solely via a floral chemical which simultaneously acts as a floral reward (e. g. methyl eugenol, raspberry ketone, or zingerone) to perform pollination. The flowers may produce attractive odours. Although absent in most species, nectar may be produced in a spur of the labellum, or on the point of the sepals, or in the septa of the ovary, the most typical position amongst the Asparagales.

In orchids that produce pollinia, pollination happens as some variant of the following sequence: when the pollinator enters into the flower, it touches a viscidium, which promptly sticks to its body, generally on the head or abdomen. While leaving the flower, it pulls the pollinium out of the anther, as it is connected to the viscidium by the caudicle or stipe. The caudicle then bends and the pollinium is moved forwards and downwards. When the pollinator enters another flower of the same species, the pollinium has taken such position that it will stick to the stigma of the second flower, just below the rostellum, pollinating it. The possessors of orchids may be able to reproduce the process with a pencil, small paintbrush, or other similar device.

Ophrys apifera is about to self-pollinate.

Some orchids mainly or totally rely on self-pollination, especially in colder regions where pollinators are particularly rare. The caudicles may dry up if the flower has not been visited by any pollinator, and the pollinia then fall directly on the stigma. Otherwise, the anther may rotate and then enter the stigma cavity of the flower (as in Holcoglossum amesianum).

The slipper orchid Paphiopedilum parishii reproduces by self-fertilization. This occurs when the anther changes from a solid to a liquid state and directly contacts the stigma surface without the aid of any pollinating agent or floral assembly.

The labellum of the Cypripedioideae is poke bonnet-shaped, and has the function of trapping visiting insects. The only exit leads to the anthers that deposit pollen on the visitor.

In some extremely specialized orchids, such as the Eurasian genus *Ophrys*, the labellum is adapted to have a colour, shape, and odour which attracts male insects via mimicry of a receptive female. Pollination happens as the insect attempts to mate with flowers.

Many neotropical orchids are pollinated by male orchid bees, which visit the flowers to gather volatile chemicals they require to synthesize pheromonal attractants. Males of such species as Euglossa imperialis or Eulaema meriana have been observed to leave their territories periodically to forage for aromatic compounds, such as cineole, to synthesize pheromone for attracting and mating with females. Each type of orchid places the pollinia on a different body part of a different species of bee, so as to enforce proper cross-pollination.

A rare achlorophyllous saprophytic orchid growing entirely underground in Australia, Rhizanthella slateri, is never exposed to light, and depends on ants and other terrestrial insects to pollinate it. Catasetum, a genus discussed briefly by Darwin, actually launches its viscid pollinia with explosive force when an insect touches a seta, knocking the pollinator off the flower.

After pollination, the sepals and petals fade and wilt, but they usually remain attached to the ovary.

Asexual Reproduction

Some species, such as Phalaenopsis, Dendrobium, and Vanda, produce offshoots or plantlets formed from one of the nodes along the stem, through the accumulation of growth hormones at that point. These shoots are known as keiki.

Fruits and Seeds

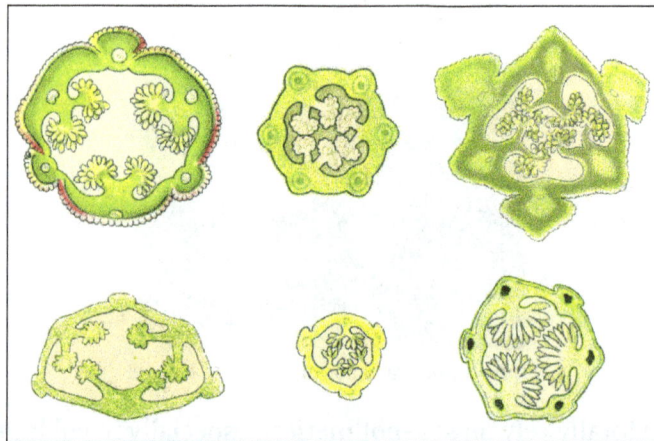

Cross-sections of orchid capsules showing the longitudinal slits.

The ovary typically develops into a capsule that is dehiscent by three or six longitudinal slits, while remaining closed at both ends.

The seeds are generally almost microscopic and very numerous, in some species over a million per capsule. After ripening, they blow off like dust particles or spores. They lack endosperm and must enter symbiotic relationships with various mycorrhizal basidiomyceteous fungi that provide them the necessary nutrients to germinate, so all orchid species are mycoheterotrophic during germination and reliant upon fungi to complete their lifecycles.

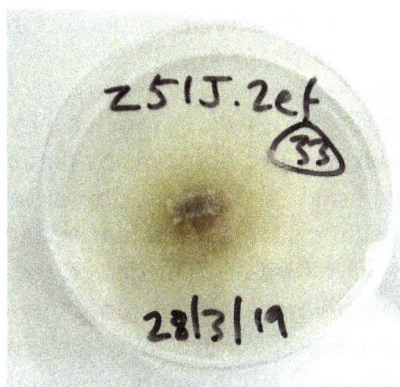

Orchid mycorrhizal fungus on agar plate,
Jodrell Laboratory, Kew Gardens.

As the chance for a seed to meet a suitable fungus is very small, only a minute fraction of all the seeds released grow into adult plants. In cultivation, germination typically takes weeks.

Horticultural techniques have been devised for germinating orchid seeds on an artificial nutrient medium, eliminating the requirement of the fungus for germination and greatly aiding the propagation of ornamental orchids. The usual medium for the sowing of orchids in artificial conditions is agar agar gel combined with a carbohydrate energy source. The carbohydrate source can be combinations of discrete sugars or can be derived from other sources such as banana, pineapple, peach, or even tomato puree or coconut water. After the preparation of the agar agar medium, it is poured into test tubes or jars which are then autoclaved (or cooked in a pressure cooker) to sterilize the medium. After cooking, the medium begins to gel as it cools.

Distribution

Orchidaceae are cosmopolitan, occurring in almost every habitat apart from glaciers. The world's richest diversity of orchid genera and species is found in the tropics, but they are also found above the Arctic Circle, in southern Patagonia, and two species of *Nematoceras* on Macquarie Island at 54° south.

The following list gives a rough overview of their distribution:

- Oceania: 50 to 70 genera,

- North America: 20 to 26 genera,

- Tropical America: 212 to 250 genera,

- Tropical Asia: 260 to 300 genera,

- Tropical Africa: 230 to 270 genera,

- Europe and temperate Asia: 40 to 60 genera.

Ecology

A majority of orchids are perennial epiphytes, which grow anchored to trees or shrubs in the tropics and subtropics. Species such as Angraecum sororium are lithophytes, growing on rocks or very rocky soil. Other orchids (including the majority of temperate Orchidaceae) are terrestrial and can be found in habitat areas such as grasslands or forest.

Some orchids, such as Neottia and Corallorhiza, lack chlorophyll, so are unable to photosynthesise. Instead, these species obtain energy and nutrients by parasitising soil fungi through the formation of orchid mycorrhizae. The fungi involved include those that form ectomycorrhizas with trees and other woody plants, parasites such as Armillaria, and saprotrophs. These orchids are known as myco-heterotrophs, but were formerly (incorrectly) described as saprophytes as it was believed they gained their nutrition by breaking down organic matter. While only a few species are achlorophyllous holoparasites, all orchids are myco-heterotrophic during germination and seedling growth, and even photosynthetic adult plants may continue to obtain carbon from their mycorrhizal fungi.

Uses

Perfumery

As decoration in a flowerpot.

A Brassolaeliocattleya ("BLC") Paradise Jewel 'Flame' hybrid orchid. Blooms of the Cattleya alliance are often used in ladies' corsages.

The scent of orchids is frequently analysed by perfumers (using headspace technology and gas-liquid chromatography/mass spectrometry) to identify potential fragrance chemicals.

Horticulture

The other important use of orchids is their cultivation for the enjoyment of the flowers. Most cultivated orchids are tropical or subtropical, but quite a few that grow in colder climates can be found on the market. Temperate species available at nurseries include Ophrys apifera (bee orchid), Gymnadenia conopsea (fragrant orchid), Anacamptis pyramidalis (pyramidal orchid) and Dactylorhiza fuchsii (common spotted orchid).

Orchids of all types have also often been sought by collectors of both species and hybrids. Many hundreds of societies and clubs worldwide have been established. These can be small, local clubs, or larger, national organisations such as the American Orchid Society. Both serve to encourage cultivation and collection of orchids, but some go further by concentrating on conservation or research.

The term "botanical orchid" loosely denotes those small-flowered, tropical orchids belonging to several genera that do not fit into the "florist" orchid category. A few of these genera contain enormous numbers of species. Some, such as Pleurothallis and Bulbophyllum, contain approximately 1700 and 2000 species, respectively, and are often extremely vegetatively diverse. The primary use of the term is among orchid hobbyists wishing to describe unusual species they grow, though it is also used to distinguish naturally occurring orchid species from horticulturally created hybrids.

New orchids are registered with the International Orchid Register, maintained by the Royal Horticultural Society.

Use as Food

Vanilla fruits drying.

The dried seed pods of one orchid genus, Vanilla (especially Vanilla planifolia), are commercially important as a flavouring in baking, for perfume manufacture and aromatherapy.

The underground tubers of terrestrial orchids [mainly Orchis mascula (early purple orchid)] are ground to a powder and used for cooking, such as in the hot beverage salep or in the Turkish frozen treat dondurma. The name salep has been claimed to come from the Arabic expression hasyu al-tha'lab, "fox testicles", but it appears more likely the name comes directly from the Arabic name sahlab. The similarity in appearance to testes naturally accounts for salep being considered an aphrodisiac.

The dried leaves of Jumellea fragrans are used to flavour rum on Reunion Island. Some saprophytic orchid species of the group Gastrodia produce potato-like tubers and were consumed as food by native peoples in Australia and can be successfully cultivated, notably Gastrodia sesamoides. Wild stands of these plants can still be found in the same areas as early aboriginal settlements, such as Ku-ring-gai Chase National Park in Australia. Aboriginal peoples located the plants in habitat by observing where bandicoots had scratched in search of the tubers after detecting the plants underground by scent.

Rose

A rose is a woody perennial flowering plant of the genus Rosa, in the family Rosaceae, or the flower it bears. There are over three hundred species and thousands of cultivars. They form a group of plants that can be erect shrubs, climbing, or trailing, with stems that are often armed with sharp prickles. Flowers vary in size and shape and are usually large and showy, in colours ranging from white through yellows and reds. Most species are native to Asia, with smaller numbers native to Europe, North America, and northwestern Africa. Species, cultivars and hybrids are all widely grown for their beauty and often are fragrant. Roses have acquired cultural significance in many societies. Rose plants range in size from compact, miniature roses, to climbers that can reach seven meters in height. Different species hybridize easily, and this has been used in the development of the wide range of garden roses.

Botany

Rose thorns are actually prickles – outgrowths of the epidermis.

The leaves are borne alternately on the stem. In most species they are 5 to 15 centimetres (2.0 to 5.9 in) long, pinnate, with (3–) 5–9 (–13) leaflets and basal stipules; the leaflets usually have a serrated margin, and often a few small prickles on the underside of the stem. Most roses are deciduous but a few (particularly from Southeast Asia) are evergreen or nearly so.

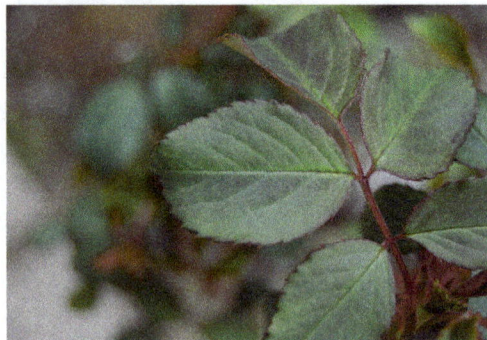

Rose leaflets.

The flowers of most species have five petals, with the exception of *Rosa sericea*, which usually has only four. Each petal is divided into two distinct lobes and is usually white or pink, though in a few species yellow or red. Beneath the petals are five sepals (or in the case of some *Rosa sericea*, four). These may be long enough to be visible when viewed from above and appear as green points alternating with the rounded petals. There are multiple superior ovaries that develop into achenes. Roses are insect-pollinated in nature.

Exterior view of rose buds.

The aggregate fruit of the rose is a berry-like structure called a rose hip. Many of the domestic cultivars do not produce hips, as the flowers are so tightly petalled that they do not provide access for pollination. The hips of most species are red, but a few (e. g. Rosa pimpinellifolia) have dark purple to black hips. Each hip comprises an outer fleshy layer, the hypanthium, which contains 5–160 "seeds" (technically dry single-seeded fruits called achenes) embedded in a matrix of fine, but stiff, hairs. Rose hips of some species, especially the dog rose (Rosa canina) and rugosa rose (Rosa rugosa), are very rich in vitamin C, among the richest sources of any plant. The hips are eaten by fruit-eating birds such as thrushes and waxwings, which then disperse the seeds in their droppings. Some birds, particularly finches, also eat the seeds.

Longitudinal section through a developing rose hip.

The sharp growths along a rose stem, though commonly called "thorns", are technically prickles, outgrowths of the epidermis (the outer layer of tissue of the stem), unlike true thorns, which are modified stems. Rose prickles are typically sickle-shaped hooks, which aid the rose in hanging onto other vegetation when growing over it. Some species such as Rosa rugosa and Rosa pimpi-nellifolia have densely packed straight prickles, probably an adaptation to reduce browsing by an-imals, but also possibly an adaptation to trap wind-blown sand and so reduce erosion and protect their roots (both of these species grow naturally on coastal sand dunes). Despite the presence of prickles, roses are frequently browsed by deer. A few species of roses have only vestigial prickles that have no points.

Species

Rosa gallica Evêque, painted by Redouté.

The genus *Rosa* is subdivided into four subgenera:

- Hulthemia (formerly Simplicifoliae, meaning "with single leaves") containing two species from southwest Asia, Rosa persica and Rosa berberifolia, which are the only roses without compound leaves or stipules.

- Hesperrhodos contains Rosa minutifolia and Rosa stellata, from North America.

- Platyrhodon with one species from east Asia, Rosa roxburghii (also known as the chestnut rose).

- Rosa (the type subgenus, sometimes incorrectly called *Eurosa*) containing all the other roses. This subgenus is subdivided into 11 sections:

  ◦ Banksianae: White and yellow flowered roses from China.

  ◦ Bracteatae: Three species, two from China and one from India.

  ◦ Caninae: Pink and white flowered species from Asia, Europe and North Africa.

  ◦ Carolinae: White, pink, and bright pink flowered species all from North America.

  ◦ Chinensis: White, pink, yellow, red and mixed-color roses from China and Burma.

  ◦ Gallicanae: Pink to crimson and striped flowered roses from western Asia and Europe.

  ◦ Gymnocarpae: One species in western North America (Rosa gymnocarpa), others in east Asia.

- ◦ Laevigatae: A single white flowered species from china.

- ◦ Pimpinellifoliae: White, pink, bright yellow, mauve and striped roses from asia and europe.

- ◦ Rosa (syn. Sect. Cinnamomeae): White, pink, lilac, mulberry and red roses from everywhere but north africa.

- ◦ Synstylae: White, pink, and crimson flowered roses from all areas.

Red roses

Uses

Roses are best known as ornamental plants grown for their flowers in the garden and sometimes indoors. They have been also used for commercial perfumery and commercial cut flower crops. Some are used as landscape plants, for hedging and for other utilitarian purposes such as game cover and slope stabilization.

Ornamental Plants

The majority of ornamental roses are hybrids that were bred for their flowers. A few, mostly species roses are grown for attractive or scented foliage (such as *Rosa glauca* and *Rosa rubiginosa*), ornamental thorns (such as *Rosa sericea*) or for their showy fruit (such as *Rosa moyesii*).

Ornamental roses have been cultivated for millennia, with the earliest known cultivation known to date from at least 500 BC in Mediterranean countries, Persia, and China. Many thousands of rose hybrids and cultivars have been bred and selected for garden use as flowering plants. Most are double-flowered with many or all of the stamens having mutated into additional petals.

In the early 19th century the Empress Josephine of France patronized the development of rose breeding at her gardens at Malmaison. As long ago as 1840 a collection numbering over one thousand different cultivars, varieties and species was possible when a rosarium was planted by Loddiges nursery for Abney Park Cemetery, an early Victorian garden cemetery and arboretum in England.

Cut Flowers

Roses are a popular crop for both domestic and commercial cut flowers. Generally they are harvested and cut when in bud, and held in refrigerated conditions until ready for display at their point of sale.

Bouquet of pink roses.

In temperate climates, cut roses are often grown in greenhouses, and in warmer countries they may also be grown under cover in order to ensure that the flowers are not damaged by weather and that pest and disease control can be carried out effectively. Significant quantities are grown in some tropical countries, and these are shipped by air to markets across the world. Some kind of roses are artificially coloured using dyed water, like rainbow roses.

Perfume

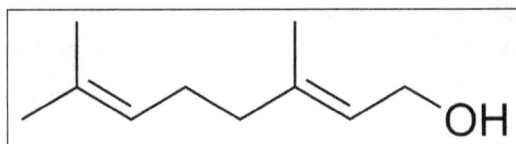
Geraniol ($C_{10}H_{18}O$).

Rose perfumes are made from rose oil (also called attar of roses), which is a mixture of volatile essential oils obtained by steam distilling the crushed petals of roses. An associated product is rose water which is used for cooking, cosmetics, medicine and religious practices. The production technique originated in Persia and then spread through Arabia and India, and more recently into eastern Europe. In Bulgaria, Iran and Germany, damask roses (*Rosa* × *damascena* 'Trigintipetala') are used. In other parts of the world *Rosa* × *centifolia* is commonly used. The oil is transparent pale yellow or yellow-grey in colour. 'Rose Absolute' is solvent-extracted with hexane and produces a darker oil, dark yellow to orange in colour. The weight of oil extracted is about one three-thousandth to one six-thousandth of the weight of the flowers; for example, about two thousand flowers are required to produce one gram of oil.

The main constituents of attar of roses are the fragrant alcohols geraniol and L-citronellol and rose camphor, an odorless solid composed of alkanes, which separates from rose oil. β-Damascenone is also a significant contributor to the scent.

Food and Drink

Rose hips are occasionally made into jam, jelly, marmalade, and soup or are brewed for tea, primarily for their high vitamin C content. They are also pressed and filtered to make rose hip syrup.

Rose hips are also used to produce rose hip seed oil, which is used in skin products and some makeup products.

Rosa canina hips.

Rose water has a very distinctive flavour and is used heavily in Middle Eastern, Persian, and South Asian cuisine—especially in sweets such as barfi, baklava, halva, gulab jamun, gumdrops, kanafeh, nougat, and Turkish delight.

Rose petals or flower buds are sometimes used to flavour ordinary tea, or combined with other herbs to make herbal teas.

In France, there is much use of rose syrup, most commonly made from an extract of rose petals. In the Indian subcontinent, Rooh Afza, a concentrated squash made with roses, is popular, as are rose-flavoured frozen desserts such as ice cream and kulfi.

Gulab jamun made with rose water.

Rose flowers are used as food, also usually as flavouring or to add their scent to food. Other minor uses include candied rose petals.

Rose creams (rose-flavoured fondant covered in chocolate, often topped with a crystallised rose petal) are a traditional English confectionery widely available from numerous producers in the UK.

Under the American Federal Food, Drug, and Cosmetic Act, there are only certain *Rosa* species, varieties, and parts are listed as generally recognized as safe (GRAS).

- Rose absolute: Rosa alba L., Rosa centifolia L., Rosa damascena Mill., Rosa gallica L., and vars. of these spp.

- Rose (otto of roses, attar of roses): Ditto.

- Rose buds: Ditto.

- Rose flowers: Ditto.

- Rose fruit (hips): Ditto.

- Rose leaves: *Rosa spp.*

Medicine

The rose hip, usually from R. canina, is used as a minor source of vitamin C. The fruits of many species have significant levels of vitamins and have been used as a food supplement. Many roses have been used in herbal and folk medicines. Rosa chinensis has long been used in Chinese traditional medicine. This and other species have been used for stomach problems, and are being investigated for controlling cancer growth. In pre-modern medicine, diarrhodon is a name given to various compounds in which red roses are an ingredient.

Hibiscus

Hibiscus is a genus of flowering plants in in the mallow family, Malvaceae. The genus is quite large, comprising several hundred species that are native to warm temperate, subtropical and tropical regions throughout the world. Member species are renowned for their large, showy flowers and those species are commonly known simply as "hibiscus", or less widely known as rose mallow. Other names include hardy hibiscus, rose of sharon, and tropical hibiscus.

The genus includes both annual and perennial herbaceous plants, as well as woody shrubs and small trees. Several species are widely cultivated as ornamental plants, notably Hibiscus syriacus and Hibiscus rosa-sinensis.

A tea made from hibiscus flowers is known by many names around the world and is served both hot and cold. The beverage is known for its red colour, tart flavour, and vitamin C content.

The leaves are alternate, ovate to lanceolate, often with a toothed or lobed margin. The flowers are large, conspicuous, trumpet-shaped, with five or more petals, colour from white to pink, red, orange, peach, yellow or purple, and from 4–18 cm broad. Flower colour in certain species, such as H. mutabilis and H. tiliaceus, changes with age. The fruit is a dry five-lobed capsule, containing several seeds in each lobe, which are released when the capsule dehisces (splits open) at maturity. It is of red and white colours. It is an example of complete flowers.

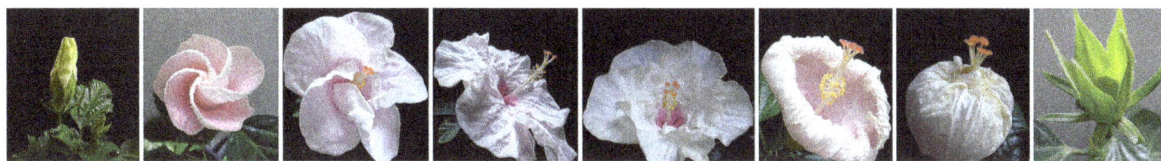
Stages in the life-cycle of a flower.

Uses

Symbolism and Culture

The red hibiscus is the flower of the Hindu goddess Kali, and appears frequently in depictions of her in the art of Bengal, India, often with the goddess and the flower merging in form. The hibiscus is used as an offering to goddess Kali and Lord Ganesha in Hindu worship.

In the Philippines, the gumamela (local name for hibiscus) is used by children as part of a bubble-making pastime. The flowers and leaves are crushed until the sticky juices come out. Hollow papaya stalks are then dipped into this and used as straws for blowing bubbles. Together with soap, hibiscus juices produce more bubbles. Also called "Tarukanga" in waray particularly in eastern samar province.

(Giant) tropical Hibiscus
rosa-sinensis 'Madonna'.

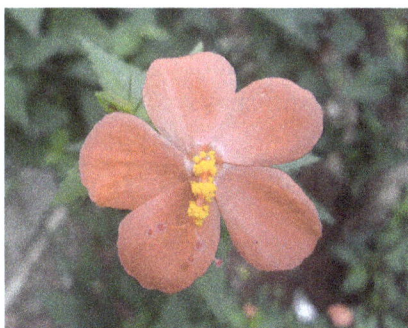
Hibiscus hirtus.

The hibiscus flower is traditionally worn by Tahitian and Hawaiian girls. If the flower is worn behind the left ear, the woman is married or has a boyfriend. If the flower is worn on the right, she is single or openly available for a relationship. The yellow hibiscus is Hawaii's state flower.

Hibiscus lobatus.

Hibiscus hispidissimus.

Nigerian author Chimamanda Ngozi Adichie named her first novel *Purple Hibiscus* after the delicate flower.

The bark of the hibiscus contains strong bast fibres that can be obtained by letting the stripped bark set in the sea to let the organic material rot away.

Yellow hibiscus.

As a National and State Symbol

The hibiscus is a national symbol of Haiti, and the national flower of nations including the Solomon Islands and Niue. *Hibiscus syriacus* is the national flower of South Korea, and *Hibiscus rosa-sinensis* is the national flower of Malaysia. *Hibiscus brackenridgei* is the state flower of Hawaii.

Landscaping

Many species are grown for their showy flowers or used as landscape shrubs, and are used to attract butterflies, bees, and hummingbirds.

Hibiscus is a very hardy, versatile plant and in tropical conditions it can enhance the beauty of any garden. Being versatile it adapts itself easily to balcony gardens in crammed urban spaces and can be easily grown in pots as a creeper or even in hanging pots. It is a perennial and flowers through the year. As it comes in a variety of colors, it's a plant which can add vibrancy to any garden.

The only infestation that gardeners need to be vigilant about is mealybug. Mealybug infestations are easy to spot as it is clearly visible as a distinct white cottony infestation on buds, leaves or even stems. To protect the plant you need to trim away the infected part, spray with water, and apply an appropriate pesticide.

Paper

One species of Hibiscus, known as kenaf (Hibiscus cannabinus), is extensively used in paper-making.

Rope and Construction

The inner bark of the sea hibiscus (Hibiscus tiliaceus), also called 'hau', is used in Polynesia for making rope, and the wood for making canoe floats. The ropes on the missionary ship Messenger of Peace were made of fibres from hibiscus trees.

Beverage

The tea made of the calyces of Hibiscus sabdariffa is known by many names in many countries around the world and is served both hot and cold. The beverage is well known for its red colour, tartness and unique flavour. Additionally, it is highly nutritious because of its vitamin C content.

It is known as bissap in West Africa, "Gul e Khatmi" in Urdu & Persian, agua de jamaica in Mexico and Central America (the flower being flor de jamaica) and Orhul in India. Some refer to it as roselle, a common name for the hibiscus flower. In Jamaica, Trinidad and many other islands in the Caribbean, the drink is known as sorrel. In Ghana, the drink is known as soobolo in one of the local languages.

In Cambodia, a cold beverage can be prepared by first steeping the petals in hot water until the colors are leached from the petals, then adding lime juice (which turns the beverage from dark brown/red to a bright red), sweeteners (sugar/honey) and finally cold water/ice cubes.

In the Arab world, hibiscus tea is known as karkadé, and is served as both a hot and a cold drink.

Food

Dried hibiscus is edible, and it is often a delicacy in Mexico. It can also be candied and used as a garnish, usually for desserts.

The roselle (Hibiscus sabdariffa) is used as a vegetable. The species Hibiscus suratensis Linn synonymous to Hibiscus aculeatus G. Don is noted in Visayas in the Philippines as being a souring ingredient for almost all local vegetables and menus. Known as labog in the Visayan area, (or labuag/sapinit in Tagalog), the species is an ingredient in cooking native chicken soup.

Hibiscus species are used as food plants by the larvae of some lepidopteran species, including Chionodes hibiscella, Hypercompe hambletoni, the nutmeg moth, and the turnip moth.

Reddish-yellow hibiscus cultivar.

Folk Medicine

Hibiscus rosa-sinensis is described as having a number of medical uses in Indian Ayurveda. It has been claimed that sour teas derived from Hibiscus sabdariffa may lower blood pressure.

Precautions and Contraindications

Pregnancy and Lactation

While the mechanism is not well understood, previous animal studies have demonstrated both an inhibitory effect of H. sabdariffa on muscle tone and the anti-fertility effects of Hibiscus rosa-sinensis, respectively. The extract of H. sabdariffa has been shown to stimulate contraction of the rat bladder and uterus; the H. rosa-sinensis extract has exhibited contraceptive effects in the form of estrogen activity in rats. These findings have not been observed in humans. The Hibiscus rosa-sinensis is also thought to have emmenagogue effects which can stimulate menstruation and, in some women, cause an abortion. Due to the documented adverse effects in animal studies and the reported pharmacological properties, the H. sabdariffa and H. rosa-sinensis are not recommended for use during pregnancy.

Yellow hibiscus cultivar.

Drug Interactions

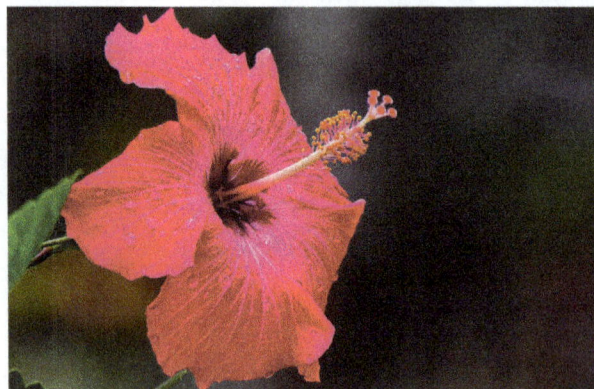

It is postulated that H. sabdariffa interacts with diclofenac, chloroquine and acetaminophen by altering the pharmacokinetics. In healthy human volunteers, the H. sabdariffa extract was found to reduce the excretion of diclofenac upon co-administration. Additionally, co-administration of

Karkade (H. sabdariffa), a common Sudanese beverage, was found to reduce chloroquine bioavailability. However, no statistically significant changes were observed in the pharmacokinetics of acetaminophen when administered with the Zobo (H. sabdariffa) drink.

Cut Flowers

Rose, hydrangea and calla wedding bouquet

Cut flowers are flowers or flower buds (often with some stem and leaf) that have been cut from the plant bearing it. It is usually removed from the plant for decorative use. Typical uses are in vase displays, wreaths and garlands. Many gardeners harvest their own cut flowers from domestic gardens, but there is a significant floral industry for cut flowers in most countries. The plants cropped vary by climate, culture and the level of wealth locally. Often the plants are raised specifically for the purpose, in field or glasshouse growing conditions. Cut flowers can also be harvested from the wild.

The cultivation and trade of flowers is a specialization in horticulture, specifically floriculture.

Garland sellers outside Banke Bihari Temple, Vrindavan, India.

Cultivation

Cut flower cultivation is intensive, usually on the basis of greenhouse monocultures, and requires large amounts of highly toxic pesticides, residues of which can often still be found in flower shops on imported flowers.

These facts have spurred the development of movements like "Slow Flowers", which propagates sustainable floriculture in the consumer country (U. S., Canada) itself.

Uses

A common use is for floristry, usually for decoration inside a house or building. Typically the cut flowers are placed in a vase. A number of similar types of decorations are used, especially in larger buildings and at events such as weddings. These are often decorated with additional foliage. In some cultures, a major use of cut flowers is for worship; this can be seen especially in south and southeast Asia.

Sometimes the flowers are picked rather than cut, without any significant leaf or stem. Such flowers may be used for wearing in hair, or in a button-hole. Masses of flowers may be used for sprinkling, in a similar way to confetti. Garlands, wreaths and bouquets are major derived and value added products.

Longevity

Once flowers are removed from the plant they continue to grow slowly, but have a diminished capability of receiving the nutrients that are vital for their survival. In most countries, cut flowers are a local crop because of their perishable nature. In India, much of the product has a shelf life of only a day. Among these are marigold flowers for garlands and temples, which are typically harvested before dawn, and discarded after use the same day.

The majority of cut flowers can be expected to last several days with proper care. This generally requires standing them in water in shade. They can be treated in various ways to increase their life. According to James C. Schmidt, a horticulturist at the University of Illinois, originally putting cut flowers in a sterilized vase is important to extending the life of the flowers. Vases can be cleaned using a household dish detergent or a combination of water and bleach. Using these disinfectants ensures that there will be less bacteria growing within the vase that could potentially cause the plant to wilt and die at a faster rate. Schmidt also claims that cutting the flowers diagonally with a sharp knife under running water ensures that they can immediately take up fresh and clean water. Re-cutting the stems periodically will ensure that there is a fresh surface from which the stems can take up water. This will allow the flowers to last even longer. Other ways to care for vase flowers includes keeping flowers away from ceiling fans and air-conditioning vents as this can lead to dehydration, keeping flowers away from fresh fruit of vegetables, using filtered water rather than tap water so as to avoid both chlorine and fluoride, and keeping flowers away from your television.

There is also a market for 'everlasting' or dried flowers, which include species such as Xerochrysum bracteatum. These can have a very long shelf life.

Additives

According to the Brooklyn Botanical Garden, different additives can be used to prolong the lives of fresh cut flowers. Experiments were performed with various substances mixed with water, including aspirin, vitamin pills, vinegar, pennies, and flower food to test their effect on cut flowers' lifespans. Each plant was placed in the same environment and the same type of plant was used in each vase.

It found that the best additive for flowers was the retailer-provided "flower food" that is usually given with a bouquet. Plants are known to thrive in an environment where there are few bacteria, plenty of food for energy, and water uptake is encouraged. Flower foods contain an acidifier that helps to adjust the water's pH. With a lower pH the water and food conducting system within the flower can work at maximum efficiency. The sugar in the food will be used by the plant as an energy source, which had been lost when the flower was cut away from its root. With these nutrients the plant will be able to fully develop. Finally, there are stem unpluggers that will make sure that the flower can easily take up water and nutrients that can later be used to take care of the needs of the rest of the plant. This combination gives the fresh cut flowers everything that they need to survive longer. When tests were carried out in St. Mary's College C. S. S. p, Rathmines, Dublin, however, results showed that glucose was more effective at prolonging the life of cut flowers than the commercial plant food. The experiment was carried out as part of the Junior Certificate Science Examination set by the Irish State Examinations Commission 2008.

Trade

The largest producers are, in order of cultivated area, China, India, and the United States. The largest importer and exporter by value is the Netherlands, which is both a grower and a redistributor of crops imported from other countries. Most of its exports go to its European neighbours.

In recent decades, with the increasing use of air freight, it has become economic for high value crops to be grown far from their point of sale; the market is usually in industrialised countries. Typical of these is the production of roses in Ecuador and Colombia, mainly for the US market, and production in Kenya and Uganda for the European market. Some countries specialise in especially high value products, such as orchids from Singapore and Thailand.

As with the production of fruit and vegetables, the industry depends on significant amounts of water, which may be collected and stored by the farm owners. The Patel Dam failure in May 2018, associated with a large Kenyan rose farm, killed dozens of people.

The total market value in most countries is considerable. It has been estimated at approximately GBP 2 billion in the United Kingdom, of the same order as that of music sales.

Hybrid Rose (Rosa Hybrida)

Varieties

Gladiator, Baby Pink, Sofia Lawrence, YCD 1, YCD 2, YCD 3 are commonly cultivated.

YCD 1   YCD 2   YCD 3

Soil and Climate

It is generally suitable for higher elevation (1500 m and above). It can also be grown in the plains under ideal condition of fertile loamy soils with salt-free irrigation water. The ideal climate for rose growing should have temperature with a minimum of 15 °C and maximum of 28 °C. Light is important factor which decides the growth. The growth is slowed by day length, i. e. > 12 hours and heavy overcast, cloudy/mist conditions. High relative humidity exposes the plant to serious fungal diseases. In tropics the ideal temperature is 25 °C – 30 °C on sunny day and on cloudy day 18 °C – 20 °C. The optimum temperature should be 15 °C – 18 °C. These temperatures are extremely difficult to find and it's therefore to compromise.

Propagation and Planting

The crop can be propagated by rooted cuttings or by budding on Briar root stocks in hills and on Edward Rose and *Rosa indica* in plains. One year old budded plants are planted in July - August at 75 cm x 75 cm spacing.

Planting of Rose

After Cultivation

The plants should be watered daily until they establish and thereafter once in a week. Pruning is done during March and October. Spray Diuran 2.5 kg a. i/ha to control weeds. Avoid spray fluid coming in contact with Rose plants.

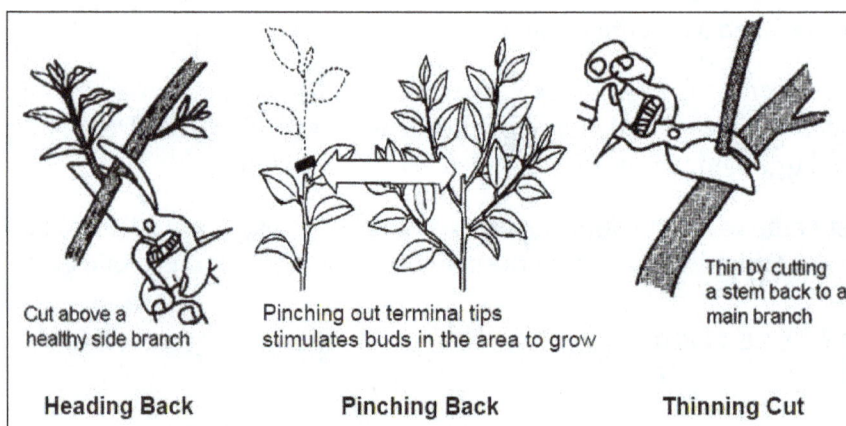

Pruning

Support of the Plants

Post is placed at internals of 3m on both sides of the bed. Along the sides of the bed, galvanized wires or plastic string are fastened at the posts at 30cm – 40cm intervals to support the plant. Between the wires across the bed, thin strings can be tied to keep the width of the beds constant.

Disbudding

Varieties produce some side buds below the center bud. These side buds have to be removed or disbudded. The disbudding must be done regularly and also as soon as possible in order to avoid large wounds in the upper leaf axil.

Dead Shoot Removal

In the old plants the dead shoot or dried shoots on plants will serve as the host for fungi. So regularly these have to be removed.

Soil Loosening on Beds

After 6 months or so, there is every chance that the soil become stony and it has to be loosened for efficient irrigation.

Bending

Leaf is a source of food for every plant. There should be balance between Source (Assimilation) and sink (Dissimilation). After planting, 2 to 3 eye buds will sprout on main branch. These sprouts will grow as branches and these branches in turn form buds. The mother shoot is bend on 2nd leaf or nearer to the crown region. The first bottom break or ground shoot will start coming from the base. These ground shoots form the basic framework for production and thereon the ground shoots should be cut at 5th five pair of leaves and medium ground shoots should be cut at 2nd or 3rd five pair of leaves.

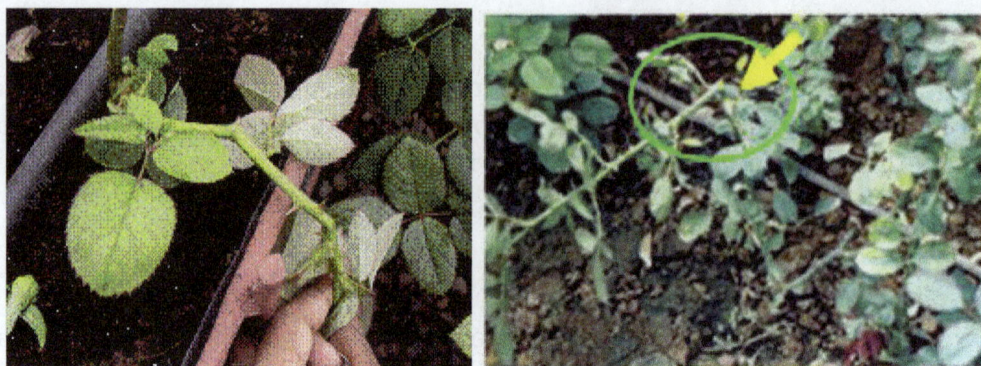
Bending.

Defoliation

The removal of leaves is known as defoliation. It is done mainly to induce certain plant species to flower or to reduce transpiration loss during periods of stress. Defoliation may be done by removal of leaves manually or by withholding water. The shoots are defoliated after pruning.

Manuring

At three months interval, apply FYM at 10 kg and 8:8:16 g NPK/plant after each pruning.

Harvest

Harvesting is done with sharp secateure at the tight bud stage when the colour is fully developed and the petals have not yet started unfolding. There should be 1-2 mature leaves (those with five leaflets) left on the plant after the flower has been cut. The reason for leaving these matures leaves is to encourage production of new strong shoots. Harvesting is done preferably during early morning hours.

Secateur for harvest.

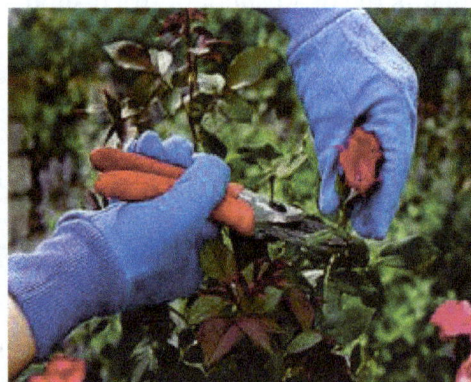
Harvesting technique.

Postharvest Handling

Roses must be placed in a bucket of water inside the polyhouse immediately after harvesting and transported to cold storage (2-4 °C). The length of time depends upon the variety and quality of the roses. The flowers are graded according to the length. It varies from 40-70 cm depending on the variety and packed in 10/12 per bunch.

Pre cooling.

Grading.

Packing.

Cut Chrysanthemum (Dendranthema Grandiflora Tzeuleu)

Varieties

Standard types	Bonfire Orange, Bonfire Yellow.
Spray types	Reagan Yellow, Reagan White, Nanako, etc.,
Climate	Cut chrysanthemums are grown under polyhouses with the following environmental conditions.
Temperature	16 - 25 °C
Relative humidity	70 - 85 %
CO_2	600 - 900 ppm
Photoperiod	Long day conditions with 13 hours light & 11 hours darkness during vegetative stage (upto 4-5 weeks from planting) and short day conditions with 10 hours light & 14 hours darkness during flower bud initiation stage.
Soil	Well drained sandy loam soil with good texture and aeration or growing medium made of 1: 1: 2 of soil, compost and cocopeat with pH of 5.5 to 6.5.

Growing Media

The growing media consists of soil, compost and coco peat in the ratio of 1:1:2. The beds are formed with 1 m width, 0.3m height and at convenient length. The soil pH must be 6.5 with 1 to 1.5 EC (Electrical Conductivity):

- Propagation: Terminal cuttings and tissue culture plants are used. Terminal cuttings are widely used for commercial cultivation. Cuttings of 5-7 cm length are taken from healthy stock plants and are induced to root by treating with IBA (1000 ppm).

- Planting: Beds of 1m width, 0.3m height and convenient length are formed. Nets (with cell size depending on the spacing adopted) are placed on the beds and planting is done.

- Spacing: 15x 15 cm (45 plants/m²) or 10 x 15 cm (67 plants/m²).

- Irrigation

  ○ Drip irrigation with 8-9 litres of water/m²/day.

  ○ Nutrition: Basal application of DAP - 50 g/m².

  ○ Weekly schedule - from 3rd week after planting.

Fertilizer	Quantity (g/m²)	
	Monday	Wednesday
19-19-19	3.0	1.0
KNO_3	3.0	1.0
CAN	2.0	1.0
Ammonium nitrate	2.0	1.0
$MgSO_4$	2.0	1.0

- Fertilizer Management

  ○ NPK @ 20:20:10 g/m² is applied through fertigation at weekly intervals.

- Growth Regulators

  ○ Alar 50 – 150 gm/100 lit water and B 9 at 8 – 25 ml/lit of water is used twice at the growing stage.

Special Practices

- Pinching

  ○ First pinching - 3 weeks after planting; 2nd pinching - 5 weeks after planting.

- Disbudding

  ○ In spray varieties, only the large apical bud is removed and the lateral buds are retained. In standard varieties, the lateral buds are removed and only apical buds are allowed to develop.

- Blindness
  - It occurs when the night temperature is too low and the days are short at the time when flower buds are forming. A rosetted type of growth is indicative of this difficulty. Center petals that fail to develop can be due to excessive heat; or in dark weather some varieties apparently lack enough food to open the flower. Chlorosis, or yellowing of the upper foliage, is generally associated with over watering, excessive fertilizer in the soil, or insects or diseases attacking the root system. Continued growth of shoots and failure to form flower buds when short days are started the mean night temperature was too low. Sunscald is prevalent on standards in flower in very warm weather. The petals turn brown and dry up.

Light Requirement

Chrysanthemum is very much influenced by light and hence photoperiod should be regulated.

Growth phase	Weeks from planting	Photoperiod
Vegetative phase	Up to 4-5 weeks from planting till the plant attains 50 to 60 cm height	Long day : 13 hrs light and 11 hrs dark
Flowering	5 -6 weeks after planting till harvest	Short day : 10 hrs light and 14 hrs dark

Lighting for chrysanthemum.

Growth regulators: Spray GA3 (50 ppm) at 30, 45 and 60 days after planting to increase flower stem length.

Weed management: Weeding and hoeing are done manually as and when required.

Plant Protection

Pests

Leaf miner: Spray Imidacloprid @ 0.5 ml/l or Acetamiprid @ 0.3 g/l.

Thrips: Spray Fipronil @ 1.0 ml/l. Keep Yellow Sticky Trap 10 nos. for 100 sq. m area.

Aphids: Spray Methyl demeton @ 2 ml/l or Monocrotophos @ 1 ml/l.

Red spider mite: Spray Abamectin 1.9 EC @ 0.5 ml/l or Propargite @ 2 ml/l.

Diseases

White Rust: Spray Azoxystrobin @ 1ml/l or Triflooxystrobin + Tebuconazole @ 0.75 g/l.

Leaf spot: Spray Macozeb @ 2g/l or Azoxystrobin @ 2 ml/l or Difenoconazole @ 0.5ml/l.

Wilt: Soil drenching with Carbendazim @ 1 g/l or Triflooxystrobin + Tebuconazole @ 0.75 g/l.

Powdery mildew: Spray Wettable Sulphur @ 2g/l or Azoxystrobin @ 1ml/l.

Harvest

Harvest Index

Standard types: Flowers are harvested when 2 - 3 rows of rays florets are perpendicular to the flower stalk.

Spray types: When 50% flowers have shown colour for distant markets; when two flowers have opened and others have shown colour for local markets.

Chrysanthemum ready to harvest.

Yield

- Standard types: 67 flower stems/m^2.

- Spray types: 260 flower stems/m^2.

Post Harvest Technology

Pulsing	Sucrose 4 % for 24 hrs (Vase life: 18 days; Control : 8.5 days)
Holding solution	BA 10 ppm + Bavistin 0.1 % + Sucrose 2 % (Vase life: 17 days; Control : 8.5 days)
Wrapping material	Polysleeves with holes (50 gauge thickness) (Shelf life: 9.25 days; Control : 6.5 days)

After harvest, the stem have to be cut at equal length (90 cm is the standard), bunched in five, putting a rubber band at the base and sliding them into a plastic sleeve and putting the bunches in plastic buckets filled with water. Early morning on the day of shipment (or night before), the bunches can be packed in boxes.

Gladiolus (Gladiolus spp)

Varieties

Tropic Sea, White Prosperity, Priscilla, Summer Sunshine, Pusa Swarnima, Jackson Ville Gold, KKL. 1, Archana, Basant Bahar, Indrani, Kalima, Kohra, Aarti, Arka Kesar, Darshan, Dhiraj, Agnirekha, Archana, Bindiya, Shree Ganesh.

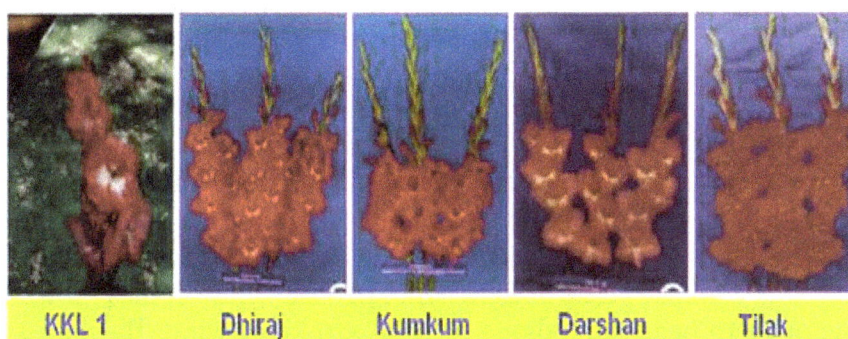

KKL 1   Dhiraj   Kumkum   Darshan   Tilak

Climate: Subtropical and temperate climatic conditions are suitable. The crop performs well under a temperature range of 27 - 30 °C. It requires full exposure to sunlight and performs well with long day conditions of 12 to 14 hour photoperiod.

Soil: Well drained sandy loam soil rich in organic matter with pH of 6 to 7.

Season

This crop requires minimum 10 hours of sunlight to over come blindness. So season should be adjusted or light substitution should be given.

Propagation: Commercial propagation is through corms. Cold storage of corms at 3 to 7 °C for 3 months or treatment with Ethrel (1000ppm) or GA3 (100ppm) or Thiourea (500 ppm) is adopted for breaking corm dormancy.

Field preparation and planting: Beds of size 6 x 2 m are prepared and corms are planted at a depth of 5 cm adopting a spacing of 40 x 25 cm (88, 888 plants/ha) or 25 x 25 cm (1, 60, 000 plants/ha).

Planting season: October for plains and March-April for hills.

Planting System

Ridges and furrows system is adopted.

Irrigation: Irrigate at 7-10 day intervals in sandy soils and at less frequent intervals in heavy soils. Irrigation should be withheld at least 4-6 weeks before lifting of corms.

Nutrition: 120 kg N, 150 kg P2O5 and 150 kg K2O per hectare is recommended, of which 60 kg N and entire dose of P2O5 and K2O is applied as basal dose. The remaining N is given in two split doses, 30 and 60 days after planting.

Manuring

Basal

N 60 kg/ha, P 150 kg/ha, K150 kg/ha.

Top Dressing

N alone is given @ 30 kg/ha during 4 leaf stage as foliar spray and 30 kg/ha during bud stage as soil application.

After Cultivation

After the corms have sprouted well, watering should be done, if necessary. When the shoots are about 20 cm high they are covered by heaping the soil up to a height of 10 to 15 cm. This enables the plants to grow erect despite high winds and rains and suppresses weed growth. Earthing up the soil is a must in case of light soils. In case where spikes grow longer or stems are not strong enough to bear the lodging or mild stroke of wind, they are supported with about 1.5 meters strong stakes. Strings instead of stakes may be used at the time of the appearance of the spikes. Strings are stretched between the stakes along the row to provide easy and adequate support.

Plant Protection

- Before storage, corms are dipped in hot water at 40 - 45 °C + fungicide (captan or thiram 2 g/lit) to control Nematode and fungal disease.
- Thrips can be controlled by methyl demeton 25 EC 2 ml/lit. or dimethoate 30 EC @ 2 ml/lit.
- Semilooper and Helicoverpa can be controlled by methyl demoton or monocrotophos @ 2 ml/lit.

Leaf Spot

Spray Carbendazim or Mancozeb 2 g/lit to control leaf spot.

Wilt

Drenching of Bavistin (0.2%) at fornight intervals controls the wilt disease.

Blight Disease

Blight disease can be controlled by spraying Mancozeb @ 0.2 %.

 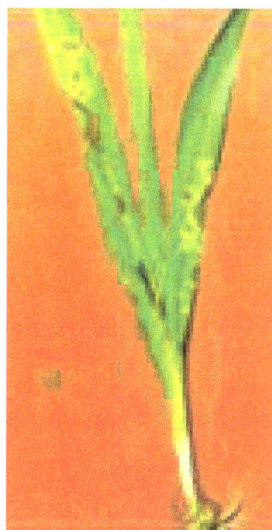

Wilt   Blight

Storage Rot

Spraying of Benomyl (0.2%) controls the storage rot.

Season of Flowering and Harvesting

When first bud shows the colour of the variety harvesting is started.

Fluoride Injury

Leaf scorch of gladioli is observed due to the presence of fluorine compounds in the atmosphere which got accumulated on the tips of leaves. The injury is associated with heavy application of super phosphate.

Special Practices

Staking: Large flowered varieties should be staked to avoid lodging.

Plant Protection

Pests

Thrips: Spray Dimethoate 30 EC @ 2ml/l or Fenitrothion 50 EC @ 2ml/l or Malathion 50 EC @ 2ml/l Semi looper and Helicoverpa: Methyl Demeton 25 EC or Monocrotophos @ 2ml/l or Dimethoate 30 EC @ 2ml/l.

Gerbera (Gerbera Jamesonii)

Varieties

TNAU Varieties: YCD-1, YCD-2.

Red	Ruby Red, Sangria
Yellow	Doni, Supernova, Mammut, Talasa
Rose	Rosalin, Salvadore
Pink	Pink Elegance, Marmara, Esmara
Orange	Carrera, Goliath, Marasol
Cream	Farida, Dalma, Snow Flake, Winter Queen

Climate

Production of quality flowers requires shade house (50%) or naturally ventilated polyhouse. Day temperature of 22-25 °C and night temperature of 12-16 °C are ideal.

Soil

Well drained, rich, light, neutral or slightly alkaline soil with pH range of 5.5 - 7.0.

Season

The crop can be cultivated throughout the year.

Propagation

Commercially propagated through division of suckers and tissue culture plants.

Field Preparation and Planting

Soil fumigation with Formaldehyde (100ml in 5l/m²) or Dazomet (30g/m²) is recommended to control soil borne pathogens (Phytophthora, Fusarium and Pythium). Raised beds of 1-2m width and 30cm height are prepared. Growing media consisting of FYM: sand: cocopeat/paddy husk (2:1:1) is ideal.

Greenhouse Cultivation of Gerbera.

Before starting gerbera cultivation, disinfection of the soil is absolutely necessary to minimize the infestation of soil borne pathogens like Phytophthora, Fusarium and Pythium which could otherwise destroy the crop completely. The beds should be drenched / fumigated with 2% formaldehyde (100 ml formalin in 5 litres of water / m² area) or methyl bromide (70 g / m²) and then covered with a plastic sheet for a minimum period of 2 to 3 days. The beds should be subsequently watered thoroughly to drain the chemicals before planting. Well developed tissue culture plants having 4-6 leaves can be planted firmly without burying the crown.

Spacing

40 x 30 cm or 30 x 30 cm

Irrigation

Drip irrigation is done once in 2 – 3 days @ 3.75 litre/drip/plant for 15 – 20 minutes. Average water requirement is about 500 – 700 ml/day/plant.

Nutrition: Fertigation is adopted from 3rd week after planting as per the following schedule.

Fertilizer	Quantity (g/500m²)
A tank (Monday, Wednesday, Friday)	
Calcium Nitrate	700
Pottasium Nitrate (13:0:46)	400
Fe EDTA / sulphate	20
B tank (Tuesday, Thursday, Saturday)	
Mono Ammonium Phosphate (12:61:0)	300
Sulphate of Potash (0:0:50)	700
Magnesium Sulphate	700
Manganese Sulphate	5
Zinc Sulphate	3
Copper Sulphate	3
Molybdenum (Sodium Molybdate)	1
Boron (Borax)	3

Manuring

Basal

- Neem cake 2.5 ton/ha.

- P - 400 g/100 sq. ft.

- MgSo4 - 0.5 kg/100 sq. ft.

Top Dressing

Calcium Ammonium Nitrate and Muriate of Potash at the ratio of 5:3 is mixed and applied at 2.5 g/plant/month.

After Cultivation

- Hand weeding is done whenever necessary.

- Remove the flower buds up to 2 months and then allow for flowering.

- Rake the soil once in 15 days to facilitate easy absorption of water, fertilizer and to provide air to the roots.

- Remove older leaves to facilitate new leaf growth and good sanitation.

Special Practices

- Leaf Pruning

  ○ Remove old leaves periodically.

Plant Protection

Pests

- Aphids: Apply Imidacloprid 17.8 % SL @ 1 ml/l or Dimethoate 30 EC @ 2 ml/l

- Whitefly: Spray Imidacloprid 17.8 % SL @ 2 ml/l or Dimethoate 30 EC @ 2 ml/l

- Thrips: Spray Fipronil @ 2 ml/l or Dimethoate 30EC @ 2 ml/l

- Red spider mite: Spray Abamectin 1.9 EC @ 0.4 ml/l or Propargite @ 1 ml/l

- Nematode: Soil application of Bacillus subtilis (BbV 57)or Pseudomonasfluorescens @ 2.5 kg/ha at the time of planting for the management of root knot nematode.

Season of Flowering and Harvesting

When flowers completely open, harvesting is done. Flower stalk is soaked in Sodium hypochloride solution (5-7 ml/lit of water) for 4-5 hours to improve vase life.

Post harvest handling: Harvesting is done when outer 2-3 rows of disc florets are perpendicular to the stalk. The heel for the stalk should be cut about 2-3 cm above the base and kept in fresh chlorinated water. Flowers should be graded and sorted out in uniform batches. Flowers packed individually in poly puches and then put in to carton boxes in two layers.

Bushiness: An abnormality characterized by numerous leaves, short petioles and small laminae, which gives some cultivars of gerbera a bushy appearance known as bushiness. Nodes are not clearly distinguished and no internode elongation is seen.

Gerbera ready for harvest.

Stem Break

It is a common post harvest disorder in cut gerberas. This is mainly caused by water imbalances. It could be ethylene controlled and associated with early senescence caused by water stress.

Yellowing and Purple Margin

Nitrogen deficiency causes yellowing and early senescence of leaves. Phosphorus deficiency causes pale yellow colour with purple margin. Increase in levels of nitrogen and phosphorus were found to promote development of suckers and improve flowering in gerbera.

Grading

Based on stem length and diameter, flowers are graded in A, B, C and D.

Yield

The crop yields 2 stems / plant / month. Harvest starts from 3rd month of planting and continued up to two years. Under open condition, 130 -160 flowers / m² / year and under greenhouse condition, 175 - 200 flowers /m²/ year can be obtained.

References

- The RHS International Lily Registrar. "Application For Registration Of A Lily Name". Royal Horticultural Society. Retrieved 6 June 2014

- flower, science: britannica.com, Retrieved 29 June, 2019

- Langston CE (January 2002). "Acute renal failure caused by lily ingestion in six cats". J. Am. Vet. Med. Assoc. 220 (1): 49–52, 36. doi:10.2460/javma.2002.220.49. PMID 12680447

- Ernst Schmidt; Mervyn Lötter; Warren McCleland (2002). Trees and shrubs of Mpumalanga and Kruger National Park. Jacana Media. p. 530. ISBN 978-1-919777-30-6

- horti-flower, horticulture: agritech.tnau.ac.in, Retrieved 30 July, 2019

- George Poinar, Jr.; Finn N. Rasmussen (2017). "Orchids from the past, with a new species in Baltic amber". Botanical Journal of the Linnean Society. 183 (3): 327–333. doi:10.1093/botlinnean/bow018

- "Rose (plant) – Britannica Online Encyclopedia". Britannica.com. 19 November 2007. Retrieved 7 December 2009

3
Flower Structure, Morphology and Development

The structure of the flower consists of a non-reproductive part, known as a perianth, and the male and female reproductive parts. This chapter closely examines the different parts which constitute the structure of flowers as well as the morphology and development of flowers to provide an extensive understanding of the subject.

The flower is the reproductive structure for flowering plants (angiosperms). Flowers are extremely diverse in size, shape, color, and so on. This makes them excellent tools for distinguishing plants. As a physiologist, you can identify a few plants from just leaf and stem, but most of the time you need a flower to identify a plant.

The flower is a short branch (stem with leaves). The nodes of this branch are very close together; the internodes are typically extremely short. The leaves of this branch are of four types: sepals, petals, stamens, and carpels. The short branch is called the receptacle and the four kinds of leaves are attached to this receptacle. In most flowers there are more than two leaves of each kind on the flower, so the leaves are in a whorled arrangement (more than two leaves per node).

The lowest whorl on the receptacle is called the calyx. It is composed of a few or many sepals. In some species, sepals are green and photosynthetic. In other species, they are showy and almost indistingishable from petals.

The next whorl on the receptacle is called the corolla. It is composed of a few or many petals. The petals are typically showy and brightly colored. They serve to attract pollinators for many species. Sometimes they are extremely fragrant. They may also exude nectar (typically at the base of the petal or in a special nectar spur) to reward the pollinator. Color patterns might include nectar guides to point the way to the reward, or a "bulls eye" target among the petals might get the flying pollinator to notice the flower.

The calyx and corolla are collectively named the perianth. This literally means "around the flower" and indicates that these whorls are the sterile parts of the flower they surround the "business" parts of the flower. A flower whose perianth is missing either of these two whorls is called incomplete. A flower with both calyx and corolla has a complete perianth.

The next whorl on the receptacle is called the androecium (literally the male household). It is composed of a few or many stamens. The stamens are specialized leaves with two distinct sections: the filament (a long stalk) and the anther (usually four sacs containing pollen grains). The function of the filament is to lift the anther to a position to effectively release pollen grains into/onto the pollinator; the filament also serves to provide the anther with xylem and phloem connections to the rest of the plant. The anther serves to produce pollen grains. The pollen grains ultimately make sperm cells; thus the idea of stamens as a male unit.

The top whorl on the receptacle, in the center of the flower, is the gynoecium (literally the female household). It is composed of a few or many carpels. The carpels may be fused together into a single, compound pistil. Note the spelling of pistil; it is not a handgun. Carpels consist of three parts, a swollen base called the ovary, a stalk called the style, and a tip called the stigma. The ovary contains a chamber called a locule, and inside the locule is one or more ovules. The ovules contain an embryo sac, and the embryo sac contains the egg. The carpel is thus a female unit.

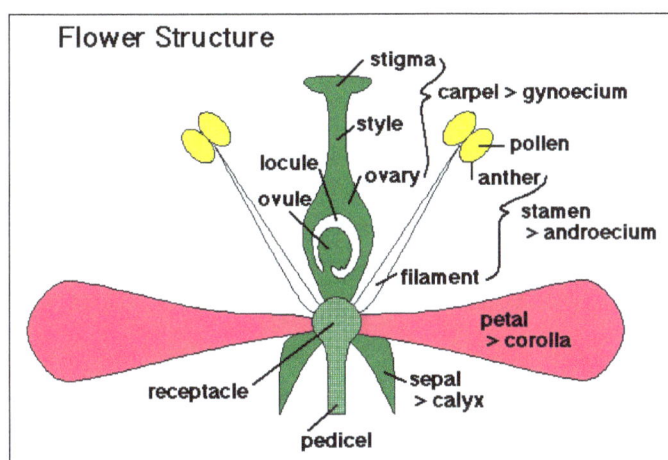

A flower can be radially or bilaterally symmetric. A sea urchin or a starfish has radial symmetry. There are several planes by which you can divide the organism (or flower) into essentially equal halves. A human is nominally bilaterally symmetric; there is only one plane of symmetry. You divide a human down the forehead, the end of the nose, the chin, the navel, and between the legs to get two equal sides (bi-lateral). Of course you are probably aware that the two sides of a human are not exactly equal. You have two eyes, but they're not the same size (to contradict the Sesame Street song). Your ears may not be attached at the same level on the sides of your head. Your heart is on one side, you liver is mostly on the other side. The appendix and spleen are found on only one side. In women, the mammary glands are not the same size. In men the testes are different sizes and are suspended differently to permit bipedal locomotion (walking on two legs). Thus a human is really asymmetric (lacking a plane of symmetry). The letter T is bilaterally symmetric. The longitudinal section of a flower shown above cannot reveal enough information to decide about symmetry, even though it might hint at least one plane of symmetry.

The flower shown above has an ovary in the superior position (the other flower parts are attached below the ovary on the receptacle). Other flowers might have the ovary sunken into the receptacle so deeply that the other flower parts appear to be attached on top of the ovary; in that case the ovary is in the inferior position.

The flower parts attached below a superior ovary are called hypogynous (below the female) while flower parts attached above an inferior ovary are called epigynous (above the female). A flower with parts attaching around the sides of the ovary are called perigynous (around the female). The ovary in that case is neither superior nor inferior in position. We generally say half-superior (or half-inferior).

A flower with both male and female parts is called perfect or bisexual or hermaphroditic. Such a flower might be able to use its own egg and sperm to reproduce, this would be called a self-pollination or a self-cross. On the other hand, such flowers might produce pollen when the stigma is not receptive to pollen, thus ensuring out-crossing or cross pollination. Sometimes the stigma can recognize a pollen grain as its own and prevent it from growing in the style; this process is called self-incompatibility.

Some plants have unisexual (imperfect) flowers: staminate (male) flowers and pistillate (female) flowers. These can be on the same plant (monoecious) or on two different plants (dioecious). A begonia is an example of a plant that has unisexual flowers but is monoecious (one household). Holly is an example of a shrub that has unisexual flowers and is dioecious (two households). Thus in planting holly, you need to take precautions. First, you need to position the female holly plant where you want a shrub that will have the red fruits on it. Second, you need to put a male holly somewhere in your landscape so that the females will be able to receive pollen to produce the fruit. You might put the fruitless male holly in a less conspicuous spot but near the females. Remember if you are at the nursery and select only shrubs with red fruits on them (a common error), those will be the last fruits you will see. You need at least one male, it "takes two to tango. " to produce fruits the next year.

Thus the range of sexuality in plants is very broad: self-crossing hermaphrodites to out-crossing hermaphrodites, to unisexual flowers, to unisexual plants. But that is really only the beginning. Plants can change sex too. Cucumbers are famous for changing from male to bisexual to female and then to parthenocarpic as they grow. What is parthenocarpic? It means literally virgin fruit. The last flowers on some cucumbers do not need to be pollinated to produce a fruit and they make the fruit on their own.

Another interesting example is the Abelmoschus (hibiscus) in the greenhouse. The flowers last only one day and are bisexual. The flower hedges its bets. In the early morning the female parts stick out beyond the stamens to be pollinated with pollen from some other plant. If this happens, fine, but if it doesn't happen by afternoon, the styles curl backwards and push the stigmas against the stamens in the flower; it is a self-pollination. By evening the flower senesces and fruits begin to develop.

Perianth

The perianth (perigonium, perigon or perigone) is the non-reproductive part of the flower, and structure that forms an envelope surrounding the sexual organs, consisting of the calyx (sepals) and the corolla (petals). In the mosses and liverworts (Marchantiophyta), the perianth is the sterile tube like tissue that surrounds the female reproductive structure (or developing sporophyte).

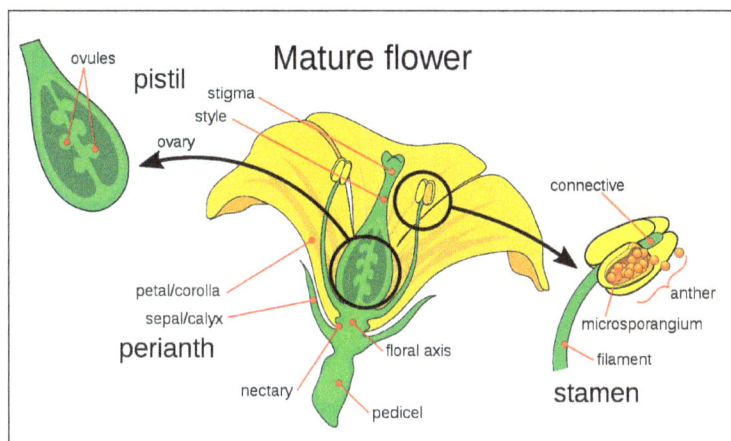

Diagram showing the parts of a mature flower. In this example
the perianth is separated into a calyx (sepals) and corolla (petals).

Flowering Plants

In flowering plants, the perianth may be described as being either dichlamydeous/heterochlamydeous in which the calyx and corolla are clearly separate, or homochlamydeous, in which they are indistinguishable (and the sepals and petals are collectively referred to as tepals). When the perianth is in two whorls, it is described as biseriate. While the calyx may be green, known as sepaloid, it may also be brightly coloured, and is then described as petaloid. When the undifferentiated tepals resemble petals, they are also referred to as "petaloid", as in petaloid monocots, orders of monocots with brightly coloured tepals. Since they include Liliales, an alternative name is lilioid monocots. The corolla and petals have a role in attracting pollinators, but this may be augmented by more specialised structures like the corona.

When the corolla consists of separate tepals the term apotepalous is used, or syntepalous if the tepals are fused to one another. The petals may be united to form a tubular corolla (gamopetalous or sympetalous). If either the petals or sepals are entirely absent, the perianth can be described as being monochlamydeous.

Types of Perianth

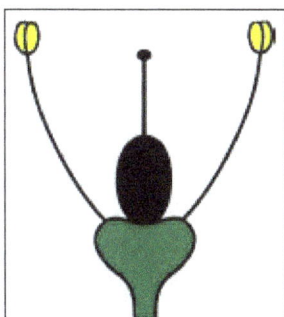

Achlamydeous floral meristem
without a corolla or calyx.

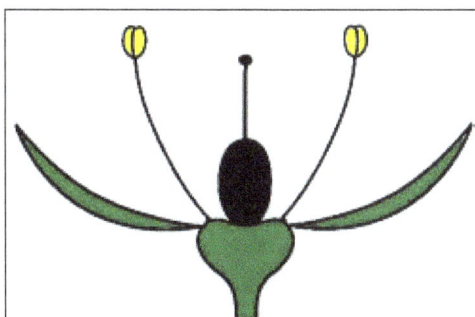

Monochlamydeous perianth with
non-petaloid calyx only.

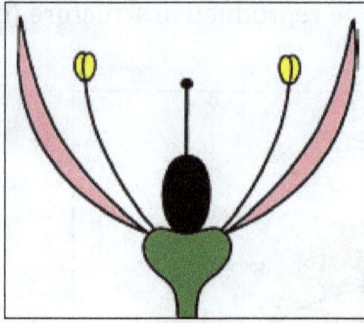

Monochlamydeous perianth with corolla only or homochlamydeous perigonium with tepals.

Dichlamydeous/heterochlamydeous perianth with separate whorls.

Both sepals and petals may have stomata and veins, even if vestigial. In some taxa, for instance some magnolias and water lilies the perianth is arranged in a spiral on nodes, rather than whorls. Flowers with spiral perianths tend to also be those with undifferentiated perianths.

Corona

Flower of Narcissus showing an outer white corolla with a central yellow corona (paraperigonium).

An additional structure in some plants (e. g. Narcissus, Passiflora (passion flower), some Hippeastrum, Liliaceae) is the corona (paraperigonium, paraperigon, or paracorolla), a ring or set of appendages of adaxial tissue arising from the corolla or the outer edge of the stamens. It is often positioned where the corolla lobes arise from the corolla tube.

Flower of Passiflora incarnata showing corona of fine appendages between petals and stamens.

The pappus of Asteraceae, considered to be a modified calyx, is also called a corona if it is shaped like a crown.

Tepal

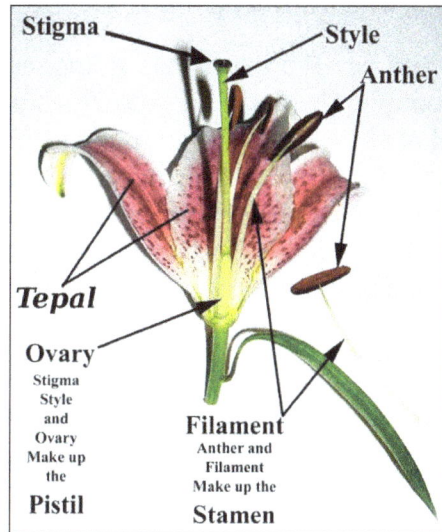

Tepal

A tepal is one of the outer parts of a flower (collectively the perianth). The term is used when these parts cannot easily be classified as either sepals or petals. This may be because the parts of the perianth are undifferentiated (i. e. of very similar appearance), as in Magnolia, or because, although it is possible to distinguish an outer whorl of sepals from an inner whorl of petals, the sepals and petals have similar appearance to one another (as in Lilium). The term was first proposed by Augustin Pyramus de Candolle in 1827 and was constructed by analogy with the terms "petal" and "sepal". (De Candolle used the term perigonium or perigone for the tepals collectively; today this term is used as a synonym for "perianth".)

A *Lilium* flower showing the six tepals: the outer
three are sepals and the inner three are petals.

Undifferentiated tepals are believed to be the ancestral condition in flowering plants. For example, *Amborella*, which is thought to have separated earliest in the evolution of flowering plants, has flowers with undifferentiated tepals. Distinct petals and sepals would therefore have arisen by differentiation, probably in response to animal pollination. In typical modern flowers, the outer or enclosing whorl of organs forms sepals, specialised for protection of the flower bud as it develops, while the inner whorl forms petals, which attract pollinators.

Tepals formed by similar sepals and petals are common in monocotyledons, particularly the "lilioid monocots". In tulips, for example, the first and second whorls both contain structures that look like petals. These are fused at the base to form one large, showy, six-parted structure (the perianth). In lilies the organs in the first whorl are separate from the second, but all look similar, thus all the showy parts are often called tepals. Where sepals and petals can in principle be distinguished, usage of the term "tepal" is not always consistent – some authors will refer to "sepals and petals" where others use "tepals" in the same context.

In some plants the flowers have no petals, and all the tepals are sepals modified to look like petals. These organs are described as petaloid, for example, the sepals of hellebores. When the undifferentiated tepals resemble petals, they are also referred to as "petaloid", as in petaloid monocots, orders of monocots with brightly coloured tepals. Since they include Liliales, an alternative name is lilioid monocots.

Sepal

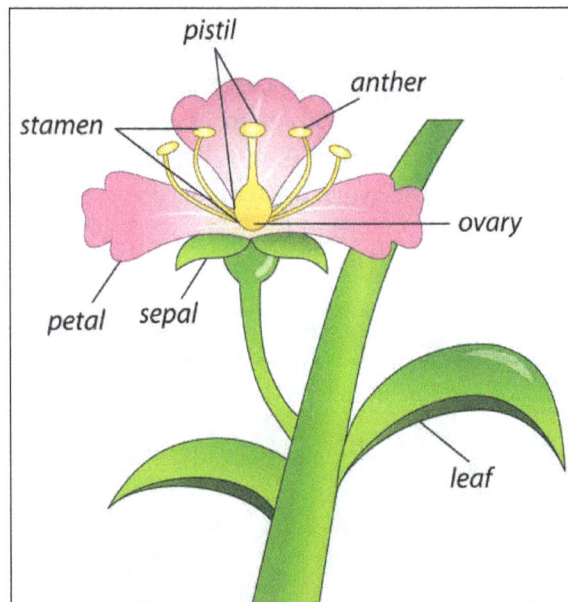

Sepals

A sepal is a part of the flower of angiosperms (flowering plants). Usually green, sepals typically function as protection for the flower in bud, and often as support for the petals when in bloom.

Collectively the sepals are called the calyx (plural calyces), the outermost whorl of parts that form a flower.

After flowering, most plants have no more use for the calyx which withers or becomes vestigial. Some plants retain a thorny calyx, either dried or live, as protection for the fruit or seeds. Examples include species of Acaena, some of the Solanaceae (for example the Tomatillo, Physalis philadelphica), and the water caltrop, Trapa natans. In some species the calyx not only persists after flowering, but instead of withering, begins to grow until it forms a bladder-like enclosure around the fruit. This is an effective protection against some kinds of birds and insects, for example in Hibiscus trionum and the Cape gooseberry. In other species, the calyx grows into an accessory fruit.

Tetramerous flower of Ludwigia
octovalvis showing petals and sepals.

Morphologically, both sepals and petals are modified leaves. The calyx (the sepals) and the corolla (the petals) are the outer sterile whorls of the flower, which together form what is known as the perianth.

The term tepal is usually applied when the parts of the perianth are difficult to distinguish, e. g. the petals and sepals share the same color, or the petals are absent and the sepals are colorful. When the undifferentiated tepals resemble petals, they are referred to as "petaloid", as in petaloid monocots, orders of monocots with brightly coloured tepals. Since they include Liliales, an alternative name is lilioid monocots. Examples of plants in which the term tepal is appropriate include genera such as Aloe and Tulipa. In contrast, genera such as Rosa and Phaseolus have well-distinguished sepals and petals.

The number of sepals in a flower is its merosity. Flower merosity is indicative of a plant's classification. The merosity of a eudicot flower is typically four or five. The merosity of a monocot or palaeodicot flower is three, or a multiple of three.

After blooming, the sepals of Hibiscus
sabdariffa expand into an edible accessory fruit.

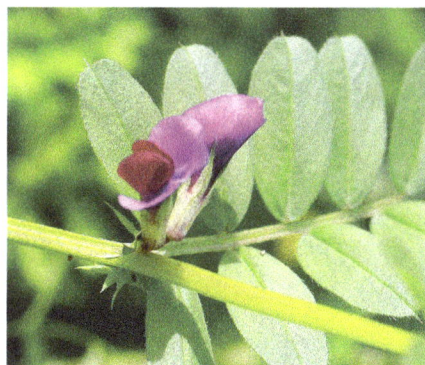

In many Fabaceae flowers, a
calyx tube surrounds the petals.

The development and form of the sepals vary considerably among flowering plants. They may be free (polysepalous) or fused together (gamosepalous). Often, the sepals are much reduced, appearing somewhat awn-like, or as scales, teeth, or ridges. Most often such structures protrude until the fruit is mature and falls off.

Examples of flowers with much reduced perianths are found among the grasses.

In some flowers, the sepals are fused towards the base, forming a calyx tube (as in the Lythraceae family, and Fabaceae). In other flowers (e. g., Rosaceae, Myrtaceae) a hypanthium includes the bases of sepals, petals, and the attachment points of the stamens.

Petal

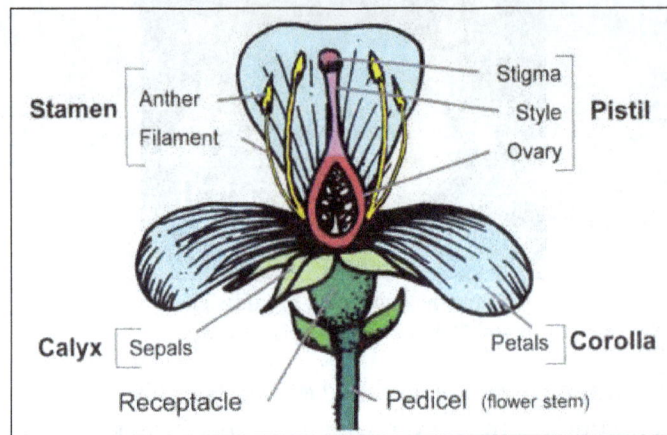

Petal

Petals are modified leaves that surround the reproductive parts of flowers. They are often brightly colored or unusually shaped to attract pollinators. Together, all of the petals of a flower are called a corolla. Petals are usually accompanied by another set of special leaves called sepals, that collectively form the calyx and lie just beneath the corolla. The calyx and the corolla together make up the perianth. When the petals and sepals of a flower are difficult to distinguish, they are collectively called tepals. Examples of plants in which the term tepal is appropriate include genera such as Aloe and Tulipa. Conversely, genera such as Rosa and Phaseolus have well-distinguished sepals and petals. When the undifferentiated tepals resemble petals, they are referred to as "petaloid", as in petaloid monocots, orders of monocots with brightly coloured tepals. Since they include Liliales, an alternative name is lilioid monocots.

Different types of petals.

Although petals are usually the most conspicuous parts of animal-pollinated flowers, wind-pollinated species, such as the grasses, either have very small petals or lack them entirely.

A Tulip's actinomorphic flower with three petals and three sepals, that collectively present a good example of an undifferentiated perianth. In this case, the word "tepals" is used.

Corolla

Apopetalous corolla.

The role of the corolla in plant evolution has been studied extensively since Charles Darwin postulated a theory of the origin of elongated corollae and corolla tubes.

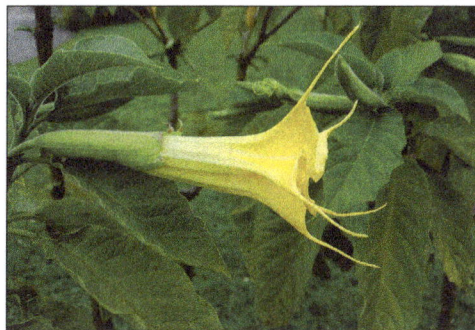

Tubular-campanulate corolla, bearing long points and emergent from tubular calyx
(Brugmansia aurea, Golden Angel's Trumpet, family Solanaceae).

A corolla of separate tepals, without fusion of individual segments, is apopetalous. If the petals are free from one another in the corolla, the plant is polypetalous or choripetalous; while if the petals are at least partially fused together, it is gamopetalous or sympetalous. In the case of fused tepals, the term is syntepalous. The corolla in some plants forms a tube.

Variations

Pelargonium peltatum, the Ivy-leaved Pelargonium : its floral structure is almost identical to that of geraniums, but it is conspicuously zygomorphic.

Petals can differ dramatically in different species. The number of petals in a flower may hold clues to a plant's classification. For example, flowers on eudicots (the largest group of dicots) most frequently have four or five petals while flowers on monocots have three or six petals, although there are many exceptions to this rule.

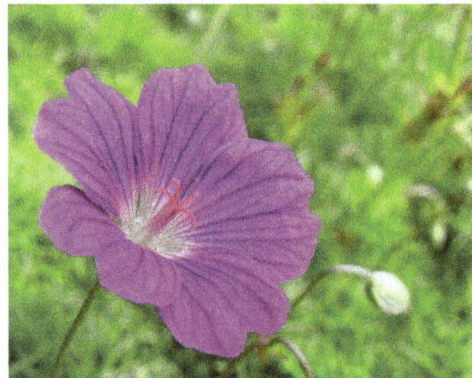

Geranium incanum, with an actinomorphic flower typical of the genus.

The petal whorl or corolla may be either radially or bilaterally symmetrical. If all of the petals are essentially identical in size and shape, the flower is said to be regular or actinomorphic (meaning "ray-formed"). Many flowers are symmetrical in only one plane (i. e., symmetry is bilateral) and are termed irregular or zygomorphic (meaning "yoke-" or "pair-formed"). In irregular flowers, other floral parts may be modified from the regular form, but the petals show the greatest deviation from radial symmetry. Examples of zygomorphic flowers may be seen in orchids and members of the pea family.

In many plants of the aster family such as the sunflower, Helianthus annuus, the circumference of the flower head is composed of ray florets. Each ray floret is anatomically an individual flower with a single large petal. Florets in the centre of the disc typically have no or very reduced petals.

In some plants such as Narcissus the lower part of the petals or tepals are fused to form a floral cup (hypanthium) above the ovary, and from which the petals proper extend.

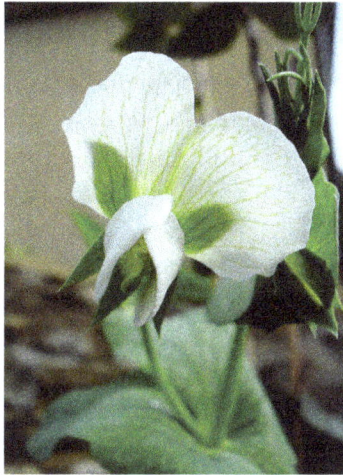

The white flower of Pisum sativum, the Garden
Pea : an example of a zygomorphic flower.

Petal often consists of two parts: the upper, broad part, similar to leaf blade, also called the blade and the lower part, narrow, similar to leaf petiole, called the claw, separated from each other at the limb. Claws are developed in petals of some flowers of the family Brassicaceae, such as Erysimum cheiri.

Narcissus pseudonarcissus, the Wild Daffodil, showing
(from bend to tip of flower) spathe, floral cup, tepals, corona

The inception and further development of petals shows a great variety of patterns. Petals of different species of plants vary greatly in colour or colour pattern, both in visible light and in ultraviolet. Such patterns often function as guides to pollinators, and are variously known as nectar guides, pollen guides, and floral guides.

Genetics

The genetics behind the formation of petals, in accordance with the ABC model of flower development, are that sepals, petals, stamens, and carpels are modified versions of each other. It appears that the mechanisms to form petals evolved very few times (perhaps only once), rather than evolving repeatedly from stamens.

Significance of Pollination

Pollination is an important step in the sexual reproduction of higher plants. Pollen is produced by the male flower or by the male organs of hermaphroditic flowers.

Pollen does not move on its own and thus requires wind or animal pollinators to disperse the pollen to the stigma (botany) of the same or nearby flowers. However, pollinators are rather selective in determining the flowers they choose to pollinate. This develops competition between flowers and as a result flowers must provide incentives to appeal to pollinators (unless the flower self-pollinates or is involved in wind pollination). Petals play a major role in competing to attract pollinators. Henceforth pollination dispersal could occur and the survival of many species of flowers could prolong.

Functions and Purposes

Petals have various functions and purposes depending on the type of plant. In general, petals operate to protect some parts of the flower and attract/repel specific pollinators.

Function

This is where the positioning of the flower petals are located on the flower is the corolla e. g. the buttercup having shiny yellow flower petals which contain guidelines amongst the petals in aiding the pollinator towards the nectar. Pollinators have the ability to determine specific flowers they wish to pollinate. Using incentives flowers draw pollinators and set up a mutual relation between each other in which case the pollinators will remember to always guard and pollinate these flowers (unless incentives are not consistently met and competition prevails).

Scent

The petals could produce different scents to allure desirable pollinators or repel undesirable pollinators. Some flowers will also mimic the scents produced by materials such as decaying meat, to attract pollinators to them.

Colour

Various colour traits are used by different petals that could attract pollinators that have poor smelling abilities, or that only come out at certain parts of the day. Some flowers are able to change the colour of their petals as a signal to mutual pollinators to approach or keep away.

Shape and Size

Furthermore, the shape and size of the flower/petals is important in selecting the type of pollinators they need. For example, large petals and flowers will attract pollinators at a large distance or that are large themselves. Collectively the scent, colour and shape of petals all play a role in attracting/repelling specific pollinators and providing suitable conditions for pollinating. Some pollinators include insects, birds, bats and the wind. In some petals, a distinction can be made between a lower narrowed, stalk-like basal part referred to as the claw, and a wider distal part referred to as the blade (or limb). Often the claw and blade are at an angle with one another.

Types of Pollination

Wind Pollination

Wind-pollinated flowers often have small, dull petals and produce little or no scent. Some of these flowers will often have no petals at all. Flowers that depend on wind pollination will produce large amounts of pollen because most of the pollen scattered by the wind tends to not reach other flowers.

Attracting Insects

Flowers have various regulatory mechanisms in order to attract insects. One such helpful mechanism is the use of colour guiding marks. Insects such as the bee or butterfly can see the ultraviolet marks which are contained on these flowers, acting as an attractive mechanism which is not visible towards the human eye. Many flowers contain a variety of shapes acting to aid with the landing of the visiting insect and also influence the insect to brush against anthers and stigmas (parts of the flower). One such example of a flower is the pohutukawa (*Metrosideros excelsa*) which acts to regulate colour within a different way. The pohutukawa contains small petals also having bright large red clusters of stamens. Another attractive mechanism for flowers is the use of scents which are highly attractive to humans. One such example is the rose. On the other hand, some flowers produce the smell of rotting meat and are attractive to insects such as flies. Darkness is another factor which flowers have adapted to as nighttime conditions limit vision and color-perception. Fragrancy can be especially useful for flowers which are pollinated at night by moths and other flying insects.

Attracting Birds

Flowers are also pollinated by birds and must be large and colorful to be visible against natural scenery. In New Zealand, such bird–pollinated native plants include: kowhai (Sophora species), flax (Phormium tenax) and kaka beak (Clianthus puniceus). Flowers adapt the mechanism on their petals to change colour in acting as a communicative mechanism for the bird to visit. An example is the tree fuchsia (Fuchsia excorticata) which are green when needing to be pollinated and turn red for the birds to stop coming and pollinating the flower.

Bat-pollinated Flowers

Flowers can be pollinated by short tailed bats. An example of this is the dactylanthus (Dactylanthus taylorii). This plant has its home under the ground acting the role of a parasite on the roots of forest trees. The dactylanthus has only its flowers pointing to the surface and the flowers lack colour but have the advantage of containing lots of nectar and a very strong scent. These act as a very useful mechanism in attracting the bat.

Inflorescence

An inflorescence is a group or cluster of flowers arranged on a stem that is composed of a main branch or a complicated arrangement of branches. Morphologically, it is the modified part of the shoot of seed plants where flowers are formed. The modifications can involve the length and the nature of the internodes and the phyllotaxis, as well as variations in the proportions, compressions, swellings, adnations, connations and reduction of main and secondary axes. Inflorescence can also be defined as the reproductive portion of a plant that bears a cluster of flowers in a specific pattern.

Aloe hereroensis, showing inflorescence with branched peduncle.

The stem holding the whole inflorescence is called a peduncle and the major axis (incorrectly referred to as the main stem) holding the flowers or more branches within the inflorescence is called the rachis. The stalk of each single flower is called a pedicel. A flower that is not part of an inflorescence is called a solitary flower and its stalk is also referred to as a peduncle. Any flower in an inflorescence may be referred to as a floret, especially when the individual flowers are particularly small and borne in a tight cluster, such as in a pseudanthium. The fruiting stage of an inflorescence is known as an infructescence.

Inflorescences may be simple (single) or complex (panicle). The rachis may be one of several types, including single, composite, umbel, spike or raceme.

General Characteristics

Inflorescences are described by many different characteristics including how the flowers are arranged on the peduncle, the blooming order of the flowers and how different clusters of flowers are grouped within it. These terms are general representations as plants in nature can have a combination of types.

Bracts

Inflorescences usually have modified shoots foliage different from the vegetative part of the plant. Considering the broadest meaning of the term, any leaf associated with an inflorescence is called a bract. A bract is usually located at the node where the main stem of the inflorescence forms, joined to the main stem of the plant, but other bracts can exist within the inflorescence itself. They serve a variety of functions which include attracting pollinators and protecting young flowers. According to the presence or absence of bracts and their characteristics we can distinguish:

- Ebracteate inflorescences: No bracts in the inflorescence.

- Bracteate inflorescences: The bracts in the inflorescence are very specialised, sometimes reduced to small scales, divided or dissected.

- Leafy inflorescences: Though often reduced in size, the bracts are unspecialised and look like the typical leaves of the plant, so that the term flowering stem is usually applied instead of inflorescence. This use is not technically correct, as, despite their 'normal' appearance, these leaves are considered, in fact, bracts, so that 'leafy inflorescence' is preferable.

- Leafy-bracted inflorescences: Intermediate between bracteate and leafy inflorescence.

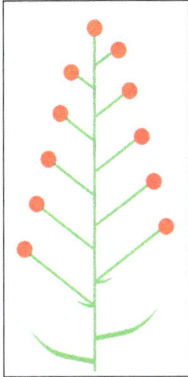
Ebracteate inflorescence.

Ebracteate of *Wisteria sinensis*.

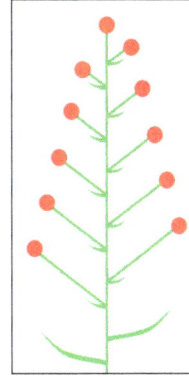
Bracteate inflorescence.

If many bracts are present and they are strictly connected to the stem, like in the family Asteraceae, the bracts might collectively be called an involucre. If the inflorescence has a second unit of bracts further up the stem, they might be called an involucel.

Bracteate inflorescence of Pedicularis verticillata.

Leafy-bracted inflorescence.

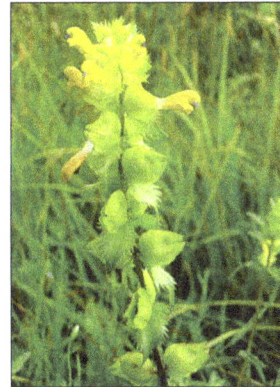
Leafy-bracted inflorescence of Rhinanthus angustifolius.

Terminal Flower

Plant organs can grow according to two different schemes, namely monopodial or racemose and sympodial or cymose. In inflorescences these two different growth patterns are called indeterminate and determinate respectively, and indicate whether a terminal flower is formed and where flowering starts within the inflorescence.

- Indeterminate inflorescence: Monopodial (racemose) growth. The terminal bud keeps growing and forming lateral flowers. A terminal flower is never formed.

- Determinate inflorescence: Sympodial (cymose) growth. The terminal bud forms a terminal flower and then dies out. Other flowers then grow from lateral buds.

Indeterminate and determinate inflorescences are sometimes referred to as open and closed inflorescences respectively. The indeterminate patterning of flowers is derived from determinate flowers. It is suggested that indeterminate flowers have a common mechanism that prevents terminal flower growth. Based on phylogenetic analyses, this mechanism arose independently multiple times in different species.

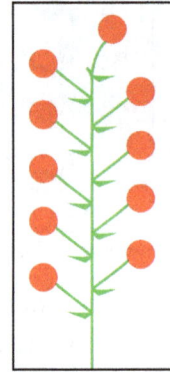

In an indeterminate inflorescence there is no true terminal flower and the stem usually has a rudimentary end. In many cases the last true flower formed by the terminal bud (subterminal flower) straightens up, appearing to be a terminal flower. Often a vestige of the terminal bud may be noticed higher on the stem.

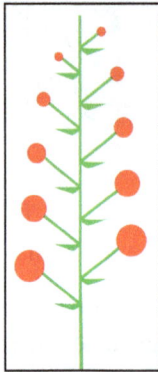

| Indeterminate inflorescence with a perfect acropetal maturation. | Indeterminate inflorescence with an acropetal maturation and lateral flower buds. | Indeterminate inflorescence with the subterminal flower to simulate the terminal one (vestige present). |

In determinate inflorescences the terminal flower is usually the first to mature (precursive development), while the others tend to mature starting from the bottom of the stem. This pattern is called acropetal maturation. When flowers start to mature from the top of the stem, maturation is basipetal, while when the central mature first, divergent.

| Determinate inflorescence with acropetal maturation. | Determinate inflorescence with basipetal maturation. | Determinate inflorescence with divergent maturation. |

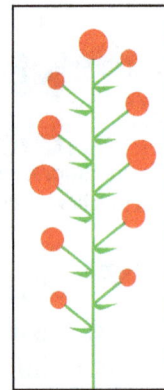

Phyllotaxis

As with leaves, flowers can be arranged on the stem according to many different patterns. Similarly arrangement of leaf in bud is called Ptyxis.

Alternate flowers.   Opposite flowers.

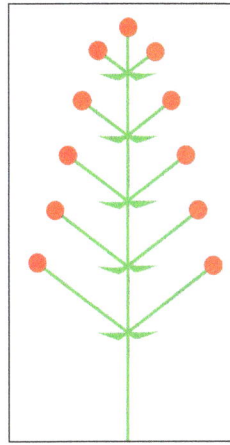

When a single or a cluster of flower(s) is located at the axil of a bract, the location of the bract in relation to the stem holding the flower(s) is indicated by the use of different terms and may be a useful diagnostic indicator.

Typical placement of bracts include:

- Some plants have bracts that subtend the inflorescence, where the flowers are on branched stalks; the bracts are not connected to the stalks holding the flowers, but are adnate or attached to the main stem (Adnate describes the fusing together of different unrelated parts. When the parts fused together are the same, they are connately joined.)

- Other plants have the bracts subtend the pedicel or peduncle of single flowers.

Metatopic placement of bracts include:

- When the bract is attached to the stem holding the flower (the pedicel or peduncle), it is said to be recaulescent; sometimes these bracts or bracteoles are highly modified and appear to be appendages of the flower calyx. Recaulescences is the fusion of the subtending leaf with the stem holding the bud or the bud itself, thus the leaf or bract is adnate to the stem of flower.

- When the formation of the bud is shifted up the stem distinctly above the subtending leaf, it is described as concaulescent.

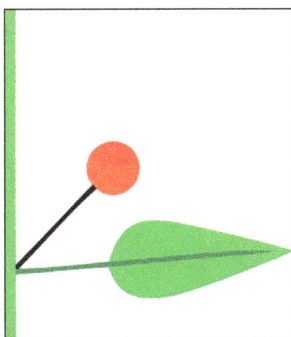

Flower and subtending bract.   Lilium martagon   Concaulescence.
                              (flower and subtending bract).

Solanum lycopersicum (concaulescence).   Recaulescence.   *Tilia cordata* (recaulescence).

Organization

There is no general consensus in defining the different inflorescences. The following is based on Focko Weberling's Morphologie der Blüten und der Blütenstände. The main groups of inflorescences are distinguished by branching. Within these groups, the most important characteristics are the intersection of the axes and different variations of the model. They may contain many flowers (pluriflor) or a few (pauciflor). Inflorescences can be simple or compound.

Simple Inflorescences

Indeterminate or Racemose

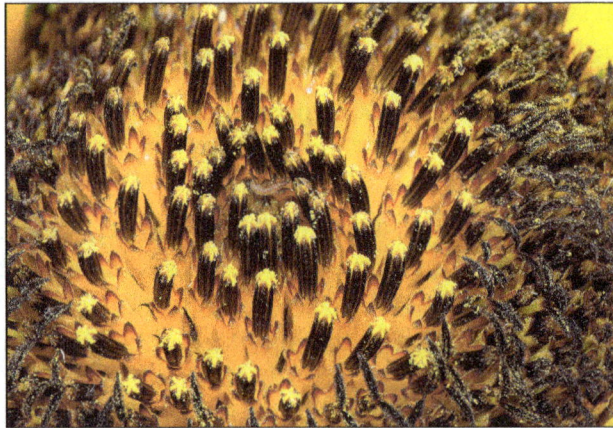

Inflorescence of sessile disc florets forming the capitulum.

Indeterminate simple inflorescences are generally called racemose. The main kind of racemose inflorescence is the raceme. The other kind of racemose inflorescences can all be derived from this one by dilation, compression, swelling or reduction of the different axes. Some passage forms between the obvious ones are commonly admitted.

- A raceme is an unbranched, indeterminate inflorescence with pedicellate (having short floral stalks) flowers along the axis.

- A spike is a type of raceme with flowers that do not have a pedicel.

- A racemose corymb is an unbranched, indeterminate inflorescence that is flat-topped or convex due to their outer pedicels which are progressively longer than inner ones.

- An umbel is a type of raceme with a short axis and multiple floral pedicels of equal length that appear to arise from a common point. It is characteristic of Umbelliferae.

- A spadix is a spike of flowers densely arranged around it, enclosed or accompanied by a highly specialised bract called a spathe. It is characteristic of the family Araceae.

- A flower head or capitulum is a very contracted raceme in which the single sessile flowers share are borne on an enlarged stem. It is characteristic of Dipsacaceae.

- A catkin or ament is a scaly, generally drooping spike or raceme. Cymose or other complex inflorescences that are superficially similar are also generally called thus.

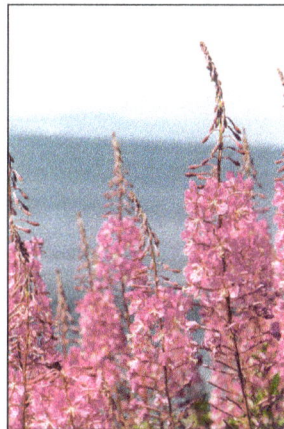

| Raceme. | Epilobium angustifolium. | Plantago media (spike). |

Determinate or Cymose

Determinate simple inflorescences are generally called cymose. The main kind of cymose inflorescence is the cyme. Cymes are further divided according to this scheme:

- Only one secondary axis: Monochasium

  ○ Secondary buds always develop on the same side of the stem: Helicoid cyme or bostryx

  ○ The Successive Pedicels are Aligned on the Same Plane: Drepanium

  ○ Secondary buds develop alternately on the stem: Scorpoid cyme

    ▪ The successive pedicels are arranged in a sort of spiral: Cincinnus (characteristic of the Boraginaceae and Commelinaceae)

    ▪ The successive pedicels follow a zig-zag path on the same plane: Rhipidium (many Iridaceae)

- Two secondary axes: Dichasial cyme

  ○ Secondary axis still dichasial: Dichasium (characteristic of Caryophyllaceae)

  ○ Secondary axis monochasia: Double scorpioid cyme or double helicoid cyme

- More than two secondary axes: Pleiochasium

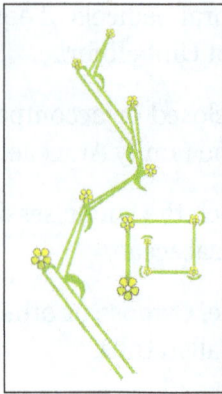

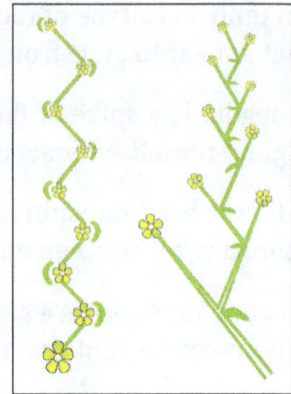

Bostryx (lateral and top view).   Gladiolus imbricatus (drepanium).   Cincinnus (lateral and top view).

A cyme can also be so compressed that it looks like an umbel. Strictly speaking this kind of inflorescence could be called umbelliform cyme, although it is normally called simply 'umbel'.

Another kind of definite simple inflorescence is the raceme-like cyme or botryoid; that is as a raceme with a terminal flower and is usually improperly called 'raceme'.

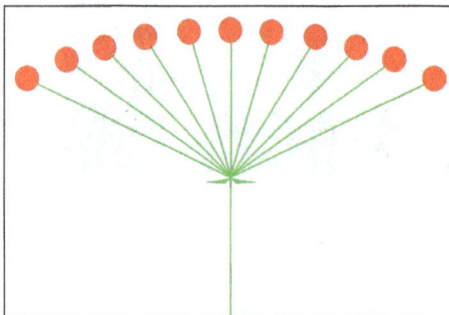

Umbelliform cyme.

Pelargonium zonale (umbelliform cyme).

Botryoid.

Berberis vernae (botryoid).

A reduced raceme or cyme that grows in the axil of a bract is called a fascicle. A verticillaster is a fascicle with the structure of a dichasium; it is common among the Lamiaceae. Many verticillasters with reduced bracts can form a spicate (spike-like) inflorescence that is commonly called a spike.

Gentiana lutea (fascicles).

Lamium orvala (verticillaster).

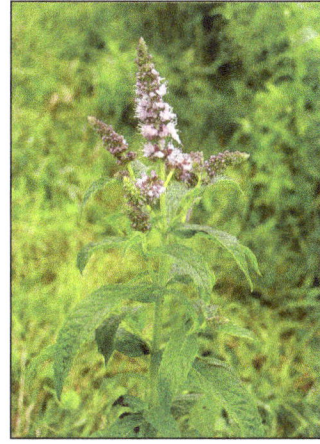
Mentha longifolia ('spike').

Compound Inflorescences

Simple inflorescences are the basis for compound inflorescences or synflorescences. The single flowers are there replaced by a simple inflorescence, which can be both a racemose or a cymose one. Compound inflorescences are composed of branched stems and can involve complicated arrangements that are difficult to trace back to the main branch.

A kind of compound inflorescence is the double inflorescence, in which the basic structure is repeated in the place of single florets. For example, a double raceme is a raceme in which the single flowers are replaced by other simple racemes; the same structure can be repeated to form triple or more complex structures.

Compound raceme inflorescences can either end with a final raceme (homoeothetic), or not (heterothetic). A compound raceme is often called a panicle. Note that this definition is very different from that given by Weberling.

Compound umbels are umbels in which the single flowers are replaced by many smaller umbels called umbellets. The stem attaching the side umbellets to the main stem is called a ray.

Homeothetic compound raceme.

Melilotus officinalis (homoeothetic compound raceme).

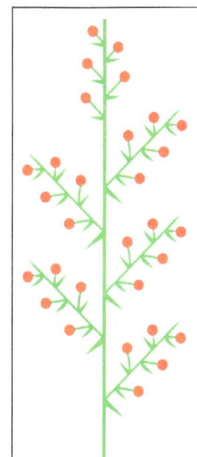
Heterothetic compound raceme.

The most common kind of definite compound inflorescence is the panicle (of Webeling, or 'panicle-like cyme'). A panicle is a definite inflorescence that is increasingly more strongly and irregularly branched from the top to the bottom and where each branching has a terminal flower.

The so-called cymose corymb is similar to a racemose corymb but has a panicle-like structure. Another type of panicle is the anthela. An anthela is a cymose corymb with the lateral flowers higher than the central ones.

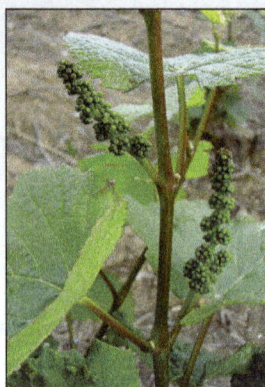

Panicle.   Vitis vinifera (panicle).   Juncus inflexus (anthela).

A raceme in which the single flowers are replaced by cymes is called a (indefinite) thyrse. The secondary cymes can be of any of the different types of dichasia and monochasia. A botryoid in which the single flowers are replaced by cymes is a definite thyrse or thyrsoid. Thyrses are often confusingly called panicles.

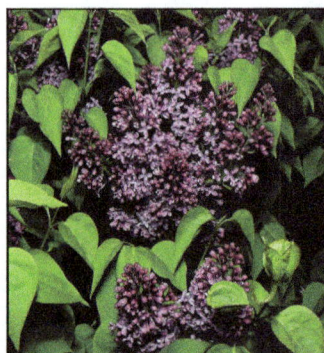

Aesculus hippocastanum.   Thyrsoid.   Syringa vulgaris.

Other combinations are possible. For example, heads or umbels may be arranged in a corymb or a panicle.

Achillea sp. (heads in a corymb).   Hedera helix (umbels in a panicle).

Other

The family Asteraceae is characterised by a highly specialised head technically called a calathid (but usually referred to as 'capitulum' or 'head'). The family Poaceae has a peculiar inflorescence of small spikes (spikelets) organised in panicles or spikes that are usually simply and improperly referred to as spike and panicle. The genus Ficus (Moraceae) has an inflorescence called syconium and the genus Euphorbia has cyathia (sing. cyathium), usually organised in umbels.

Matricaria chamomilla (calathid).   *Ficus carica* (syconium).   *Euphorbia tridentata* (cyathium).

Development and Patterning

Genetic Basis

Genes that shape inflorescence development have been studied at great length in *Arabidopsis*. *LEAFY* (LFY) is a gene that promotes floral meristem identity, regulating inflorescence development in *Arabidopsis*. Any alterations in timing of LFY expression can cause formation of different inflorescences in the plant. Genes similar in function to LFY include *APETALA1* (AP1). Mutations in LFY, AP1, and similar promoting genes can cause conversion of flowers into shoots. In contrast to LEAFY, genes like *terminal flower* (TFL) support the activity of an inhibitor that prevents flowers from growing on the inflorescence apex (flower primordium initiation), maintaining inflorescence meristem identity. Both types of genes help shape flower development in accordance with the ABC model of flower development. Studies have been recently conducted or are ongoing for homologs of these genes in other flower species.

Environmental Influences

Inflorescence-feeding insect herbivores shape inflorescences by reducing lifetime fitness (how much flowering occurs), seed production by the inflorescences, and plant density, among other traits. In the absence of this herbivory, inflorescences usually produce more flower heads and seeds. Temperature can also variably shape inflorescence development. High temperatures can impair the proper development of flower buds or delay bud development in certain species, while in others, an increase in temperature can hasten inflorescence development.

Meristems and Inflorescence Architecture

The shift from the vegetative to reproductive phase of a flower involves the development of an inflorescence meristem that generates floral meristems. Plant inflorescence architecture depends on which meristems becomes flowers and which become shoots. Consequently, genes that regulate floral meristem identity play major roles in determining inflorescence architecture because their expression domain will direct where the plant's flowers are formed.

On a larger scale, inflorescence architecture affects the quality and quantity of offspring from selfing and outcrossing, as the architecture can influence pollination success. For example, *Asclepias* inflorescences have been shown to have an upper size limit, shaped by self-pollination levels due to crosses between inflorescences on the same plant or between flowers on the same inflorescence. In *Aesculus sylvatica*, it has been shown that the most common inflorescence sizes are correlated with the highest fruit production as well.

Flower Reproductive Morphology

Plant reproductive morphology is the study of the physical form and structure (the morphology) of those parts of plants directly or indirectly concerned with sexual reproduction.

Among all living organisms, flowers, which are the reproductive structures of angiosperms, are the most varied physically and show a correspondingly great diversity in methods of reproduction. Plants that are not flowering plants (green algae, mosses, liverworts, hornworts, ferns and gymnosperms such as conifers) also have complex interplays between morphological adaptation and environmental factors in their sexual reproduction. The breeding system, or how the sperm from one plant fertilizes the ovum of another, depends on the reproductive morphology, and is the single most important determinant of the genetic structure of nonclonal plant populations. Christian Konrad Sprengel studied the reproduction of flowering plants and for the first time it was understood that the pollination process involved both biotic and abiotic interactions. Charles Darwin's theories of natural selection utilized this work to build his theory of evolution, which includes analysis of the coevolution of flowers and their insect pollinators.

Close-up of a flower of Schlumbergera (Christmas or Holiday Cactus), showing part of the gynoecium (the stigma and part of the style is visible) and the stamens that surround it.

Use of Sexual Terminology

Plants have complex lifecycles involving alternation of generations. One generation, the sporophyte, gives rise to the next generation asexually via spores. Spores may be identical isospores or come in different sizes (microspores and megaspores), but strictly speaking, spores and sporophytes are neither male nor female because they do not produce gametes. The alternate generation, the gametophyte, produces gametes, eggs and/or sperm. A gametophyte can be monoicous (bisexual), producing both eggs and sperm or dioicous (unisexual), either female (producing eggs) or male (producing sperm).

Dioicous gametophytes of the liverwort Marchantia polymorpha. In this species, gametes are produced on different plants on umbrella-shaped gametophores with different morphologies. The radiating arms of female gameteophores (left) protect archegonia that produce eggs. Male gametophores (right) are topped with antheridia that produce sperm.

In the bryophytes (liverworts, mosses and hornworts), the sexual gametophyte is the dominant generation. In ferns and seed plants (including cycads, conifers, flowering plants, etc.) the sporophyte is the dominant generation. The obvious visible plant, whether a small herb or a large tree, is the sporophyte, and the gametophyte is very small. In seed plants, each female gametophyte, and the spore that gives rise to it, is hidden within the sporophyte and is entirely dependent on it for nutrition. Each male gametophyte typically consists of from two to four cells enclosed within the protective wall of a pollen grain.

The sporophyte of a flowering plant is often described using sexual terms (e. g. "female" or "male") based on the sexuality of the gametophyte it gives rise to. For example, a sporophyte that produces spores that give rise only to male gametophytes may be described as "male", even though the sporophyte itself is asexual, producing only spores. Similarly, flowers produced by the sporophyte may be described as "unisexual" or "bisexual", meaning that they give rise to either one sex of gametophyte or both sexes of gametophyte.

Flowering Plants

Basic Flower Morphology

The flower is the characteristic structure concerned with sexual reproduction in flowering plants (angiosperms). Flowers vary enormously in their construction (morphology). A "complete" flower, like that of *Ranunculus glaberrimus* shown in the figure, has a calyx of outer sepals and a corolla of inner petals. The sepals and petals together form the perianth. Next inwards there are numerous stamens, which produce pollen grains, each containing a microscopic male gametophyte. Stamens may be called the "male" parts of a flower and collectively form the androecium. Finally in the middle there are carpels, which at maturity contain one or more ovules, and within each ovule is a tiny female gametophyte. Carpels may be called the "female" parts of a flower and collectively form the gynoecium.

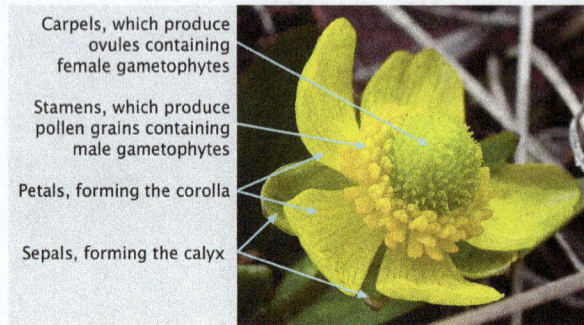

Carpels, which produce
ovules containing
female gametophytes

Stamens, which produce
pollen grains containing
male gametophytes

Petals, forming the corolla

Sepals, forming the calyx

Flower of *Ranunculus glaberrimus*.

Each carpel in Ranunculus species is an achene that produces one ovule, which when fertilized becomes a seed. If the carpel contains more than one seed, as in Eranthis hyemalis, it is called a follicle. Two or more carpels may be fused together to varying degrees and the entire structure, including the fused styles and stigmas may be called a pistil. The lower part of the pistil, where the ovules are produced, is called the ovary. It may be divided into chambers (locules) corresponding to the separate carpels.

Variations

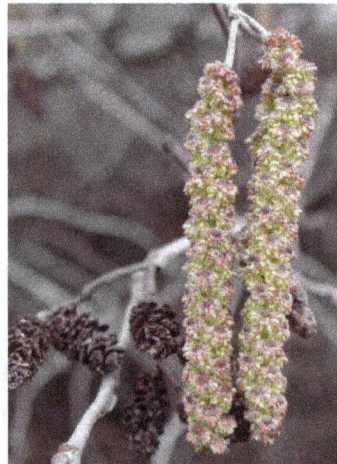

Alnus serrulata has unisexual flowers and is monoecious.

A "perfect" flower has both stamens and carpels, and may be described as "bisexual" or "hermaphroditic". A "unisexual" flower is one in which either the stamens or the carpels are missing, vestigial or otherwise non-functional. Each flower is either "staminate" (having only functional stamens) and thus "male", or "carpellate" (or "pistillate") (having only functional carpels) and thus "female". If separate staminate and carpellate flowers are always found on the same plant, the species is called monoecious. If separate staminate and carpellate flowers are always found on different plants, the species is called dioecious. A 1995 study found that about 6% of angiosperm species are dioecious, and that 7% of genera contain some dioecious species.

Members of the birch family (Betulaceae) are examples of monoecious plants with unisexual flowers. A mature alder tree (Alnus species) produces long catkins containing only male flowers, each with four stamens and a minute perianth, and separate stalked groups of female flowers, each without a perianth.

Ilex aquifolium is dioecious: (above) shoot with flowers from male plant; (top right) male flower enlarged, showing stamens with pollen and reduced, sterile stigma; (below) shoot with flowers from female plant; (lower right) female flower enlarged, showing stigma and reduced, sterile stamens (staminodes) with no pollen.

Most hollies (members of the genus Ilex) are dioecious. Each plant produces either functionally male flowers or functionally female flowers. In Ilex aquifolium, the common European holly, both kinds of flower have four sepals and four white petals; male flowers have four stamens, female flowers usually have four non-functional reduced stamens and a four-celled ovary. Since only female plants are able to set fruit and produce berries, this has consequences for gardeners. Amborella represents the first known group of flowering plants to separate from their common ancestor. It too is dioecious; at any one time, each plant produces either flowers with functional stamens but no carpels, or flowers with a few non-functional stamens and a number of fully functional carpels. However, Amborella plants may change their "sex" over time. In one study, five cuttings from a male plant produced only male flowers when they first flowered, but at their second flowering three switched to producing female flowers.

In extreme cases, all of the parts present in a complete flower may be missing, so long as at least one carpel or one stamen is present. This situation is reached in the female flowers of duckweeds (Lemna), which comprise a single carpel, and in the male flowers of spurges (Euphorbia) which comprise a single stamen.

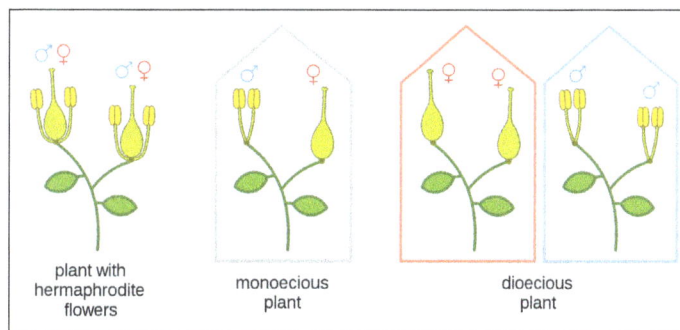

The basic cases of sexuality of flowering plants.

A species such as Fraxinus excelsior, the common ash of Europe, demonstrates one possible kind of variation. Ash flowers are wind-pollinated and lack petals and sepals. Structurally, the flowers may be bisexual, consisting of two stamens and an ovary, or may be male (staminate), lacking a functional ovary, or female (carpellate), lacking functional stamens. Different forms may occur on the same tree, or on different trees. The Asteraceae (sunflower family), with close to 22, 000 species worldwide, have highly modified inflorescences made up of flowers (florets) collected together into tightly packed heads. Heads may have florets of one sexual morphology – all bisexual, all carpellate or all staminate (when they are called homogamous), or may have mixtures of two or more sexual forms (heterogamous). Thus goatsbeards (Tragopogon species) have heads of bisexual florets, like other members of the tribe Cichorieae, whereas marigolds (Calendula species) generally have heads with the outer florets bisexual and the inner florets staminate (male).

Like Amborella, some plants undergo sex-switching. For example, Arisaema triphyllum (Jack-in-the-pulpit) expresses sexual differences at different stages of growth: smaller plants produce all or mostly male flowers; as plants grow larger over the years the male flowers are replaced by more female flowers on the same plant. Arisaema triphyllum thus covers a multitude of sexual conditions in its lifetime: nonsexual juvenile plants, young plants that are all male, larger plants with a mix of both male and female flowers, and large plants that have mostly female flowers. Other plant populations have plants that produce more male flowers early in the year and as plants bloom later in the growing season they produce more female flowers.

Terminology

The complexity of the morphology of flowers and its variation within populations has led to a rich terminology.

- Androdioecious: Having male flowers on some plants, bisexual ones on others.

- Androecious: Having only male flowers (the male of a dioecious population); producing pollen but no seed.

- Androgynous: Bisexual.

- Androgynomonoecious: Having male, female, and bisexual flowers on the same plant, also called trimonoecious.

- Andromonoecious: Having both bisexual and male flowers on the same plant.

- Bisexual: Each flower of each individual has both male and female structures, i. e. it combines both sexes in one structure. Flowers of this kind are called perfect, having both stamens and carpels. Other terms used for this condition are androgynous, hermaphroditic, monoclinous and synoecious.

- Dichogamous: Having sexes developing at different times; producing pollen when the stigmas are not receptive, either protandrous or protogynous. This promotes outcrossing by limiting self-pollination. Some dichogamous plants have bisexual flowers, others have unisexual flowers.

- Diclinous.

- Dioecious: Having either only male or only female flowers. No individual plant of the population produces both pollen and ovules.

- Gynodioecious: Having hermaphrodite flowers and female flowers on separate plants.

- Gynoecious: Having only female flowers (the female of a dioecious population); producing seed but not pollen.

- Gynomonoecious: Having both bisexual and female flowers on the same plant.

- Hermaphroditic.

- Imperfect: (of flowers) Having some parts that are normally present not developed, e. g. lacking stamens.

- Monoclinous.

- Monoecious: In the commoner narrow sense of the term, it refers to plants with unisexual flowers which occur on the same individual. In the broad sense of the term, it also includes plants with bisexual flowers. Individuals bearing separate flowers of both sexes at the same time are called simultaneously or synchronously monoecious and individuals that bear flowers of one sex at one time are called consecutively monoecious.

- Perfect (of flowers).

- Polygamodioecious: Mostly dioecious, but with either a few flowers of the opposite sex or a few bisexual flowers on the same plant.

- Polygamomonoecious: Polygamous. Or, mostly monoecious, but also partly polygamous.

- Polygamous: Having male, female, and bisexual flowers on the same plant. Also called polygamomonoecious or trimonoecious. Or, with bisexual and at least one of male and female flowers on the same plant.

- Protandrous: (of dichogamous plants) having male parts of flowers developed before female parts, e. g. having flowers that function first as male and then change to female or producing pollen before the stigmas of the same plant are receptive. (Protoandrous is also used.)

- Protogynous: (of dichogamous plants) having female parts of flowers developed before male parts, e. g. having flowers that function first as female and then change to male or producing pollen after the stigmas of the same plant are receptive.

- Subandroecious: Having mostly male flowers, with a few female or bisexual flowers.

- Subdioecious: Having some individuals in otherwise dioecious populations with flowers that are not clearly male or female. The population produces normally male or female plants with unisexual flowers, but some plants may have bisexual flowers, some both male and female flowers, and others some combination thereof, such as female and bisexual flowers. The condition is thought to represent a transition between bisexuality and dioecy.

- Subgynoecious: Having mostly female flowers, with a few male or bisexual flowers.

- Synoecious.

- Trimonoecious: Polygamous and androgynomonoecious.

- Trioecious: Polygamous.

- Unisexual: Having either functionally male or functionally female flowers. This condition is also called diclinous, incomplete or imperfect.

Outcrossing

Outcrossing, cross-fertilization or allogamy, in which offspring are formed by the fusion of the gametes of two different plants, is the most common mode of reproduction among higher plants. About 55% of higher plant species reproduce in this way. An additional 7% are partially cross-fertilizing and partially self-fertilizing (autogamy). About 15% produce gametes but are principally self-fertilizing with significant out-crossing lacking. Only about 8% of higher plant species reproduce exclusively by non-sexual means. These include plants that reproduce vegetatively by runners or bulbils, or which produce seeds without embryo fertilization (apomixis). The selective advantage of outcrossing appears to be the masking of deleterious recessive mutations.

The primary mechanism used by flowering plants to ensure outcrossing involves a genetic mechanism known as self-incompatibility. Various aspects of floral morphology promote allogamy. In plants with bisexual flowers, the anthers and carpels may mature at different times, plants being protandrous (with the anthers maturing first) or protogynous (with the carpels mature first). Monoecious species, with unisexual flowers on the same plant, may produce male and female flowers at different times.

Dioecy, the condition of having unisexual flowers on different plants, necessarily results in outcrossing, and might thus be thought to have evolved for this purpose. However, "dioecy has proven difficult to explain simply as an outbreeding mechanism in plants that lack self-incompatibility". Resource-allocation constraints may be important in the evolution of dioecy, for example, with wind-pollination, separate male flowers arranged in a catkin that vibrates in the wind may provide better pollen dispersal. In climbing plants, rapid upward growth may be essential, and resource allocation to fruit production may be incompatible with rapid growth, thus giving an advantage to delayed production of female flowers. Dioecy has evolved separately in many different lineages, and monoecy in the plant lineage correlates with the evolution of dioecy, suggesting that dioecy can evolve more readily from plants that already produce separate male and female flowers.

Pollen

Tulip anther with many grains of pollen.

Closeup image of a cactus flower and its stamens.

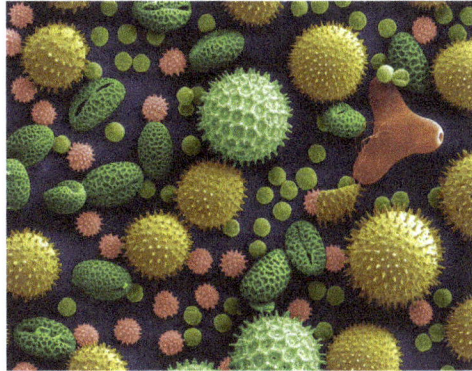

Scanning electron microscope image (500x magnification) of pollen grains from a variety of common plants: sunflower (Helianthus annuus), morning glory (Ipomoea purpurea), prairie hollyhock (Sidalcea malviflora), oriental lily (Lilium auratum), evening primrose (Oenothera fruticosa), and castor bean (Ricinus communis).

Pollen is a fine to coarse powdery substance comprising pollen grains which are male microgametophytes of seed plants, which produce male gametes (sperm cells). Pollen grains have a hard coat made of sporopollenin that protects the gametophytes during the process of their movement from the stamens to the pistil of flowering plants, or from the male cone to the female cone of coniferous plants. If pollen lands on a compatible pistil or female cone, it germinates, producing a pollen tube that transfers the sperm to the ovule containing the female gametophyte. Individual pollen grains are small enough to require magnification to see detail. The study of pollen is called palynology and is highly useful in paleoecology, paleontology, archaeology, and forensics. Pollen in plants is used for transferring haploid male genetic material from the anther of a single flower to the stigma of another in cross-pollination. In a case of self-pollination, this process takes place from the anther of a flower to the stigma of the same flower.

Pollen is infrequently used as food and food supplement. Because of agricultural practices, it is often contaminated by agricultural pesticides.

Structure and Formation

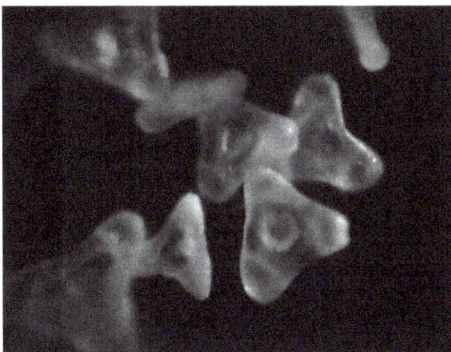

Triporate pollen of Oenothera speciosa.

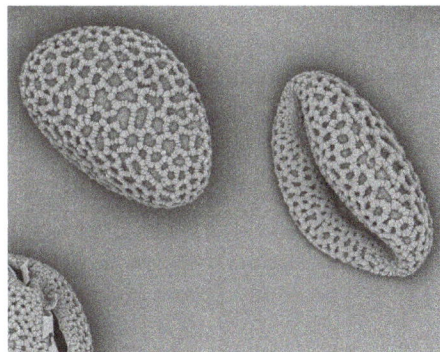

Pollen of Lilium auratum showing single sulcus (monosulcate).

Pollen itself is not the male gamete. Each pollen grain contains vegetative (non-reproductive) cells (only a single cell in most flowering plants but several in other seed plants) and a generative (reproductive) cell. In flowering plants the vegetative tube cell produces the pollen tube, and the generative cell divides to form the two sperm cells.

Arabis pollen has three colpi and prominent surface structure.

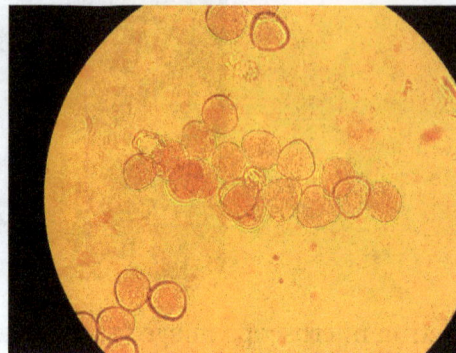

Apple pollen under microscopy.

Formation

Pollen is produced in the microsporangia in the male cone of a conifer or other gymnosperm or in the anthers of an angiosperm flower. Pollen grains come in a wide variety of shapes, sizes, and surface markings characteristic of the species. Pollen grains of pines, firs, and spruces are winged. The smallest pollen grain, that of the forget-me-not (*Myosotis* spp.), is around 6 µm (0.006 mm) in diameter. Wind-borne pollen grains can be as large as about 90–100 µm.

In angiosperms, during flower development the anther is composed of a mass of cells that appear undifferentiated, except for a partially differentiated dermis. As the flower develops, four groups of sporogenous cells form within the anther. The fertile sporogenous cells are surrounded by layers of sterile cells that grow into the wall of the pollen sac. Some of the cells grow into nutritive cells that supply nutrition for the microspores that form by meiotic division from the sporogenous cells.

In a process called microsporogenesis, four haploid microspores are produced from each diploid sporogenous cell (microsporocyte, pollen mother cell or meiocyte), after meiotic division. After the formation of the four microspores, which are contained by callose walls, the development of the pollen grain walls begins. The callose wall is broken down by an enzyme called callase and the freed pollen grains grow in size and develop their characteristic shape and form a resistant outer wall called the exine and an inner wall called the intine. The exine is what is preserved in the fossil record. Two basic types of microsporogenesis are recognised, simultaneous and successive. In simultaneous microsporogenesis meiotic steps I and II are completed prior to cytokinesis, whereas in successive microsporogenesis cytokinesis follows. While there may be a continuum with intermediate forms, the type of microsporogenesis has systematic significance. The predominant form amongst the monocots is successive, but there are important exceptions.

During microgametogenesis, the unicellular microspores undergo mitosis and develop into mature microgametophytes containing the gametes. In some flowering plants, germination of the pollen grain may begin even before it leaves the microsporangium, with the generative cell forming the two sperm cells.

Structure

Except in the case of some submerged aquatic plants, the mature pollen grain has a double wall. The vegetative and generative cells are surrounded by a thin delicate wall of unaltered cellulose called the endospore or intine, and a tough resistant outer cuticularized wall composed largely of sporopollenin called the exospore or exine. The exine often bears spines or warts, or is variously sculptured, and the character of the markings is often of value for identifying genus, species, or even cultivar or individual. The spines may be less than a micron in length (spinulus, plural spinuli) referred to as spinulose (scabrate), or longer than a micron (echina, echinae) referred to as echinate. Various terms also describe the sculpturing such as reticulate, a net like appearance consisting of elements (murus, muri) separated from each other by a lumen (plural lumina). These reticulations may also be referred to as brochi.

The pollen wall protects the sperm while the pollen grain is moving from the anther to the stigma; it protects the vital genetic material from drying out and solar radiation. The pollen grain surface is covered with waxes and proteins, which are held in place by structures called sculpture elements on the surface of the grain. The outer pollen wall, which prevents the pollen grain from shrinking and crushing the genetic material during desiccation, is composed of two layers. These two layers are the tectum and the foot layer, which is just above the intine. The tectum and foot layer are separated by a region called the columella, which is composed of strengthening rods. The outer wall is constructed with a resistant biopolymer called sporopollenin.

Pollen apertures are regions of the pollen wall that may involve exine thinning or a significant reduction in exine thickness. They allow shrinking and swelling of the grain caused by changes in moisture content. Elongated apertures or furrows in the pollen grain are called colpi (singular: colpus) or sulci (singular: sulcus). Apertures that are more circular are called pores. Colpi, sulci and pores are major features in the identification of classes of pollen. Pollen may be referred to as inaperturate (apertures absent) or aperturate (apertures present). The aperture may have a lid (operculum), hence is described as operculate. However the term inaperturate covers a wide range of morphological types, such as functionally inaperturate (cryptoaperturate) and omniaperturate. Inaperaturate pollen grains often have thin walls, which facilitates pollen tube germination at any position. Terms such as uniaperturate and triaperturate refer to the number of apertures present (one and three respectively).

The orientation of furrows (relative to the original tetrad of microspores) classifies the pollen as sulcate or colpate. Sulcate pollen has a furrow across the middle of what was the outer face when the pollen grain was in its tetrad. If the pollen has only a single sulcus, it is described as monosulcate, has two sulci, as bisulcate, or more, as polysulcate. Colpate pollen has furrows other than across the middle of the outer faces. Eudicots have pollen with three colpi (tricolpate) or with shapes that are evolutionarily derived from tricolpate pollen. The evolutionary trend in plants has been from monosulcate to polycolpate or polyporate pollen.

Additionally, gymnosperm pollen grains often have air bladders, or vesicles, called sacci. The sacci are not actually balloons, but are sponge-like, and increase the buoyancy of the pollen grain and help keep it aloft in the wind, as most gymnosperms are anemophilous. Pollen can be monosaccate, (containing one saccus) or bisaccate (containing two sacci). Modern pine, spruce, and yellow-wood trees all produce saccate pollen.

Pollination

European honey bee carrying pollen in
a pollen basket back to the hive.

Marmalade hoverfly, pollen on its face
and legs, sitting on a rockrose.

The transfer of pollen grains to the female reproductive structure (pistil in angiosperms) is called pollination. This transfer can be mediated by the wind, in which case the plant is described as anemophilous (literally wind-loving). Anemophilous plants typically produce great quantities of very lightweight pollen grains, sometimes with air-sacs. Non-flowering seed plants (e. g., pine trees) are characteristically anemophilous. Anemophilous flowering plants generally have inconspicuous flowers. Entomophilous (literally insect-loving) plants produce pollen that is relatively heavy, sticky and protein-rich, for dispersal by insect pollinators attracted to their flowers. Many insects and some mites are specialized to feed on pollen, and are called palynivores.

Diadasia bee straddles flower carpels while
visiting yellow Opuntia engelmannii cactus.

In non-flowering seed plants, pollen germinates in the pollen chamber, located beneath the micropyle, underneath the integuments of the ovule. A pollen tube is produced, which grows into the nucellus to provide nutrients for the developing sperm cells. Sperm cells of Pinophyta and Gnetophyta are without flagella, and are carried by the pollen tube, while those of Cycadophyta and Ginkgophyta have many flagella.

When placed on the stigma of a flowering plant, under favorable circumstances, a pollen grain puts forth a pollen tube, which grows down the tissue of the style to the ovary, and makes its way along the placenta, guided by projections or hairs, to the micropyle of an ovule. The nucleus of the tube cell has meanwhile passed into the tube, as does also the generative nucleus, which divides (if it hasn't already) to form two sperm cells. The sperm cells are carried to their destination in the tip of the pollen tube. Double-strand breaks in DNA that arise during pollen tube growth appear to be efficiently repaired in the generative cell that carries the male genomic information to be passed on to the next plant generation. However, the vegetative cell that is responsible for tube elongation appears to lack this DNA repair capability.

In the Fossil Record

Pollen's sporopollenin outer sheath affords it some resistance to the rigours of the fossilisation process that destroy weaker objects; it is also produced in huge quantities. There is an extensive fossil record of pollen grains, often disassociated from their parent plant. The discipline of palynology is devoted to the study of pollen, which can be used both for biostratigraphy and to gain information about the abundance and variety of plants alive — which can itself yield important information about paleoclimates. Also, pollen analysis has been widely used for reconstructing past changes in vegetation and their associated drivers. Pollen is first found in the fossil record in the late Devonian period, but at that time it is indistinguishable from spores. It increases in abundance until the present day.

Allergy to Pollen

Nasal allergy to pollen is called pollinosis, and allergy specifically to grass pollen is called hay fever. Generally, pollens that cause allergies are those of anemophilous plants (pollen is dispersed by air currents.) Such plants produce large quantities of lightweight pollen (because wind dispersal is random and the likelihood of one pollen grain landing on another flower is small), which can be carried for great distances and are easily inhaled, bringing it into contact with the sensitive nasal passages.

Pollen allergies are common in polar and temperate climate zones, where production of pollen is seasonal. In the tropics pollen production varies less by the season, and allergic reactions less. In northern Europe, common pollens for allergies are those of birch and alder, and in late summer wormwood and different forms of hay. Grass pollen is also associated with asthma exacerbations in some people, a phenomenon termed thunderstorm asthma.

In the US, people often mistakenly blame the conspicuous goldenrod flower for allergies. Since this plant is entomophilous (its pollen is dispersed by animals), its heavy, sticky pollen does not become independently airborne. Most late summer and fall pollen allergies are probably caused by ragweed, a widespread anemophilous plant.

Arizona was once regarded as a haven for people with pollen allergies, although several ragweed species grow in the desert. However, as suburbs grew and people began establishing irrigated lawns and gardens, more irritating species of ragweed gained a foothold and Arizona lost its claim of freedom from hay fever.

Anemophilous spring blooming plants such as oak, birch, hickory, pecan, and early summer grasses may also induce pollen allergies. Most cultivated plants with showy flowers are entomophilous and do not cause pollen allergies.

The number of people in the United States affected by hay fever is between 20 and 40 million, and such allergy has proven to be the most frequent allergic response in the nation. There are certain evidential suggestions pointing out hay fever and similar allergies to be of hereditary origin. Individuals who suffer from eczema or are asthmatic tend to be more susceptible to developing long-term hay fever.

In Denmark, decades of rising temperatures cause pollen to appear earlier and in greater numbers, as well as introduction of new species such as ragweed.

The most efficient way to handle a pollen allergy is by preventing contact with the material. Individuals carrying the ailment may at first believe that they have a simple summer cold, but hay fever becomes more evident when the apparent cold does not disappear. The confirmation of hay fever can be obtained after examination by a general physician.

Treatment

Antihistamines are effective at treating mild cases of pollinosis, this type of non-prescribed drugs includes loratadine, cetirizine and chlorpheniramine. They do not prevent the discharge of histamine, but it has been proven that they do prevent a part of the chain reaction activated by this biogenic amine, which considerably lowers hay fever symptoms.

Decongestants can be administered in different ways such as tablets and nasal sprays.

Allergy immunotherapy (AIT) treatment involves administering doses of allergens to accustom the body to pollen, thereby inducing specific long-term tolerance. Allergy immunotherapy can be administered orally (as sublingual tablets or sublingual drops), or by injections under the skin (subcutaneous). Discovered by Leonard Noon and John Freeman in 1911, allergy immunotherapy represents the only causative treatment for respiratory allergies.

Nutrition

Most major classes of predatory and parasitic arthropods contain species that eat pollen, despite the common perception that bees are the primary pollen-consuming arthropod group. Many other Hymenoptera other than bees consume pollen as adults, though only a small number feed on pollen as larvae (including some ant larvae). Spiders are normally considered carnivores but pollen is an important source of food for several species, particularly for spiderlings, which catch pollen on their webs. It is not clear how spiderlings manage to eat pollen however, since their mouths are not large enough to consume pollen grains. Some predatory mites also feed on pollen, with some species being able to subsist solely on pollen, such as Euseius tularensis, which feeds on the pollen of dozens of plant species. Members of some beetle families such as Mordellidae and Melyridae feed almost exclusively on pollen as adults, while various lineages within larger families such as Curculionidae, Chrysomelidae, Cerambycidae, and Scarabaeidae are pollen specialists even though most members of their families are not (e. g., only 36 of 40, 000 species of ground beetles, which are typically predatory, have been shown to eat pollen—but this is thought to be a severe underestimate as the feeding habits are only known for 1, 000 species). Similarly, Ladybird beetles mainly eat insects, but many species also eat pollen, as either part or all of their diet. Hemiptera are mostly herbivores or omnivores but pollen feeding is known (and has only been well studied in the Anthocoridae). Many adult flies, especially Syrphidae, feed on pollen, and three UK syrphid species feed strictly on pollen (syrphids, like all flies, cannot eat pollen directly due to the structure of their mouthparts, but can consume pollen contents that are dissolved in a fluid). Some species of fungus, including Fomes fomentarius, are able to break down grains of pollen as a secondary nutrition source that is particularly high in nitrogen. Pollen may be valuable diet supplement for detritivores, providing them with nutrients needed for growth, development and maturation. It was suggested that obtaining nutrients from pollen, deposited on the forest floor during periods of pollen rains, allows fungi to decompose nutritionally scarce litter.

Some species of Heliconius butterflies consume pollen as adults, which appears to be a valuable nutrient source, and these species are more distasteful to predators than the non-pollen consuming species.

Although bats, butterflies and hummingbirds are not pollen eaters *per se*, their consumption of nectar in flowers is an important aspect of the pollination process.

In Humans

Bee pollen for human consumption is marketed as a food ingredient and as a dietary supplement. The largest constituent is carbohydrates, with protein content ranging from 7 to 35 percent depending on the plant species collected by bees.

Honey produced by bees from natural sources contains pollen derived p-coumaric acid, an antioxidant and natural bactericide that is also present in a wide variety of plants and plant-derived food products.

The U. S. Food and Drug Administration (FDA) has not found any harmful effects of bee pollen consumption, except from the usual allergies. However, FDA does not allow bee pollen marketers in the United States to make health claims about their produce, as no scientific basis for these has ever been proven. Furthermore, there are possible dangers not only from allergic reactions but also from contaminants such as pesticides and from fungi and bacteria growth related to poor storage procedures. A manufacturers's claim that pollen collecting helps the bee colonies is also controversial.

Pine pollen is traditionally consumed in Korea as an ingredient in sweets and beverages.

Parasites

The growing industries in pollen harvesting for human and bee consumption rely on harvesting pollen baskets from honey bees as they return to their hives using a pollen trap. When this pollen has been tested for parasites, it has been found that a multitude of pollinator viruses and eukaryotic parasites are present in the pollen. It is currently unclear if the parasites are introduced by the bee that collected the pollen or if it is from contamination to the flower. Though this is not likely to pose a risk to humans, it is a major issue for the bumblebee rearing industry that relies on thousands of tonnes of honey bee collected pollen per year. Several sterilization methods have been employed, though no method has been 100% effective at sterilizing, without reducing the nutritional value, of the pollen.

Pollen Grain Staining

For agricultural research purposes, assessing the viability of pollen grains can be necessary and illuminating. A very common, efficient method to do so is known as Alexander's stain. This differential stain consists of ethanol, malachite green, distilled water, glycerol, phenol, chloral hydrate, acid fuchsin, orange, and glacial acetic acid. In angiosperms and gymnosperms non-aborted pollen grain will appear red or pink, and aborted pollen grains will appear blue or slightly green.

Tapetum

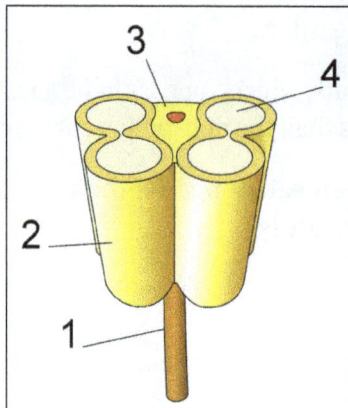

Schematic of anther (1: Filament 2: Theca 3: Connective 4: Pollen sac or Microsporangium).

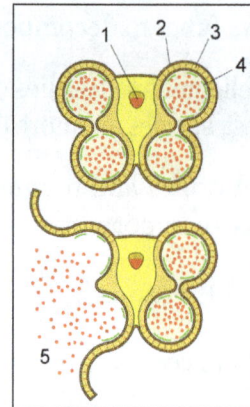

Section of anther, showing dehiscence and release of pollen (1: Vascular bundle 2: Epidermis 3: Fibrous layer 4: Tapetum 5: Pollen.

The tapetum is a specialised layer of nutritive cells found within the anther, of flowering plants, where it is located between the sporangenous tissue and the anther wall. Tapetum is important for the nutrition and development of pollen grains, as well as a source of precursors for the pollen coat. The cells are usually bigger and normally have more than one nucleus per cell. As the sporogenous cells undergo mitosis, the nuclei of tapetal cells also divide. Sometimes, this mitosis is not normal due to which many cells of mature tapetum become multinucleate. Sometimes polyploidy and polyteny can also be seen. The unusually large nuclear constitution of the tapetum helps it in providing nutrients and regulatory molecules to the forming pollen grains. The following processes are responsible for this:

- Endomitosis,

- Normal mitosis not followed by cytokinesis,

- Formation of restitution nuclei,

- Endoreduplication.

Tapetum helps in pollenwall formation, transportation of nutrients to inner side of anther, synthesis of callase enzyme for separation of microspore tetrads.

Types of Tapetum

Two main tapetum types are recognised, secretory (glandular) and plasmodial (amoeboid). In the secretory type a layer of tapetal cells remains around the anther locule, while in the plasmodial type the tapetal cell walls dissolve and their protoplasts fuse to form a multinucleate plasmodium. A third, less common type, the invasive non-syncytial tapetum has been described in Canna, where the tapetal cell walls break down to invade the anther locule but do not fuse to form a plasmodium.

Amongst the monocots Acorales, the first branching clade has a secretory tapetum, while the other alismatid clade, Alismatales are predominantly plasmodial. Amongst the late branching clades, the lilioid monocots are nearly all secretory while the commelinid monocots are diverse with respect to tapetal pattern.

Female Reproductive Parts of Flower

The main female reproductive part of a flower is called the pistil. Located in the center of the flower, the pistil holds the ovules, or what will become seeds, after pollination. It's easy to identify the pistil by its three distinctive parts. Coming out of the center of the pistil is a tube called the style. On the bottom end, the style attaches to the ovary, the part of the plant that produces the ovules. Attached to the top of the style is the stigma, a sticky knob that catches pollen.

Gynoecium

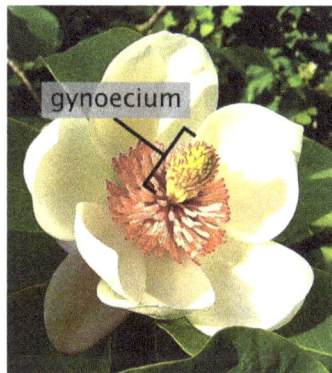

Flower of Magnolia × wieseneri showing the many pistils
making up the gynoecium in the middle of the flower.

Gynoecium is most commonly used as a collective term for the parts of a flower that produce ovules and ultimately develop into the fruit and seeds. The gynoecium is the innermost whorl of a flower; it consists of (one or more) pistils and is typically surrounded by the pollen-producing reproductive organs, the stamens, collectively called the androecium. The gynoecium is often referred to as the "female" portion of the flower, although rather than directly producing female gametes (i. e. egg cells), the gynoecium produces megaspores, each of which develops into a female gametophyte which then produces egg cells.

Hippeastrum flowers showing
stamens, style and stigma.

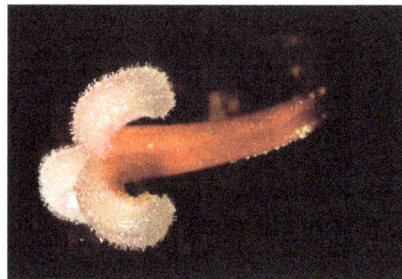

Hippeastrum stigmas and style.

Moss plants with gynoecia, clusters of
archegonia at the apex of each shoot.

The term gynoecium is also used by botanists to refer to a cluster of archegonia and any associated modified leaves or stems present on a gametophyte shoot in mosses, liverworts, and hornworts. The corresponding terms for the male parts of those plants are clusters of antheridia within the androecium. Flowers that bear a gynoecium but no stamens are called pistillate or carpellate. Flowers lacking a gynoecium are called staminate.

The gynoecium is often referred to as female because it gives rise to female (egg-producing) gametophytes; however, strictly speaking sporophytes do not have a sex, only gametophytes do.

Gynoecium development and arrangement is important in systematic research and identification of angiosperms, but can be the most challenging of the floral parts to interpret.

Pistils

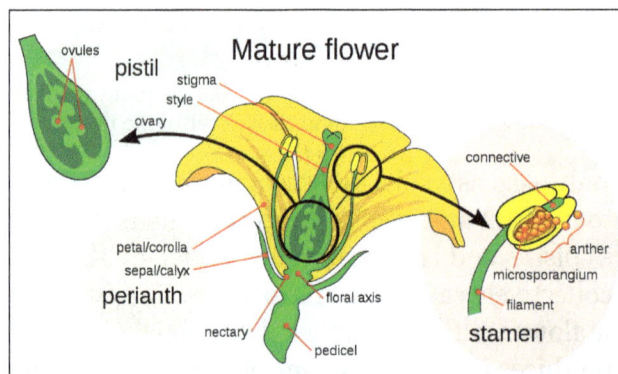

A syncarpous gynoecium.

The gynoecium (whether composed of a single carpel or multiple "fused" carpels) is typically made up of an ovary, style, and stigma as in the center of the flower. The gynoecium may consist of one or more separate pistils. A pistil typically consists of an expanded basal portion called the ovary, an elongated section called a style and an apical structure that receives pollen called a stigma.

- The ovary is the enlarged basal portion which contains placentas, ridges of tissue bearing one or more ovules (integumented megasporangia). The placentas and/or ovule(s) may be born on the gynoecial appendages or less frequently on the floral apex. The chamber in which the ovules develop is called a locule (or sometimes cell).

- The style is a pillar-like stalk through which pollen tubes grow to reach the ovary. Some flowers such as Tulipa do not have a distinct style, and the stigma sits directly on the ovary. The style is a hollow tube in some plants such as lilies, or has transmitting tissue through which the pollen tubes grow.

- The stigma is usually found at the tip of the style, the portion of the carpels that receives pollen (male gametophytes). It is commonly sticky or feathery to capture pollen.

A sterile pistil in a male flower is referred to as a pistillode.

Carpels

The pistils of a flower are considered to be composed of carpels. A carpel is the female reproductive part of the flower, interpreted as modified leaves bearing structures called ovules, inside which the egg cells ultimately form. A pistil may consist of one carpel, with its ovary, style and stigma, or several carpels may be joined together with a single ovary, the whole unit called a pistil. The gynoecium may consist of one or more uni-carpellate (with one carpel) pistils, or of one multi-carpellate pistil. The number of carpels is described by terms such as tricarpellate (three carpels).

Carpels are thought to be phylogenetically derived from ovule-bearing leaves or leaf homologues (megasporophylls), which evolved to form a closed structure containing the ovules. This structure is typically rolled and fused along the margin.

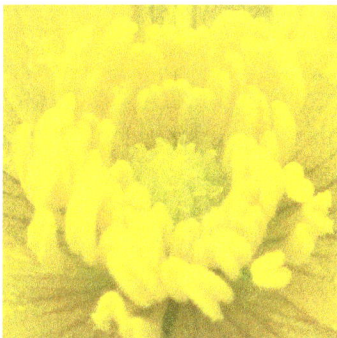

Centre of a *Ranunculus repens* (creeping buttercup) showing multiple unfused carpels surrounded by longer stamens.

Cross-section through the ovary of *Narcissus* showing multiple connate carpels (a compound pistil) fused along the placental line where the ovules form in each locule.

Although many flowers satisfy the above definition of a carpel, there are also flowers that do not have carpels according to this definition because in these flowers the ovule(s), although enclosed, are borne directly on the shoot apex. Different remedies have been suggested for this problem. An easy remedy that applies to most cases is to redefine the carpel as an appendage that encloses ovule(s) and may or may not bear them.

Types

If a gynoecium has a single carpel, it is called monocarpous. If a gynoecium has multiple, distinct (free, unfused) carpels, it is apocarpous. If a gynoecium has multiple carpels "fused" into a single structure, it is syncarpous. A syncarpous gynoecium can sometimes appear very much like a monocarpous gynoecium.

Comparison of gynoecium terminology using carpel and pistil			
Gynoecium composition	Carpel terminology	Pistil terminology	Examples
Single carpel	Monocarpous (unicarpellate) gynoecium	A pistil (simple)	Avocado (Persea sp.), most legumes (Fabaceae)
Multiple distinct ("unfused") carpels	Apocarpous (choricarpous) gynoecium	Pistils (simple)	Strawberry (Fragaria sp.), Buttercup (Ranunculus sp.)
Multiple connate ("fused") carpels	Syncarpous gynoecium	A pistil (compound)	Tulip (Tulipa sp.), most flowers

The degree of connation ("fusion") in a syncarpous gynoecium can vary. The carpels may be "fused" only at their bases, but retain separate styles and stigmas. The carpels may be "fused" entirely, except for retaining separate stigmas. Sometimes (e. g., Apocynaceae) carpels are fused by their styles or stigmas but possess distinct ovaries. In a syncarpous gynoecium, the "fused" ovaries of the constituent carpels may be referred to collectively as a single compound ovary. It can be a challenge to determine how many carpels fused to form a syncarpous gynoecium. If the styles and stigmas are distinct, they can usually be counted to determine the number of carpels. Within the compound ovary, the carpels may have distinct locules divided by walls called septa. If a syncarpous gynoecium has a single style and stigma and a single locule in the ovary, it may be necessary to examine how the ovules are attached. Each carpel will usually have a distinct line of placentation where the ovules are attached.

Pistil Development

Pistils begin as small primordia on a floral apical meristem, forming later than, and closer to the (floral) apex than sepal, petal and stamen primordia. Morphological and molecular studies of pistil ontogeny reveal that carpels are most likely homologous to leaves.

A carpel has a similar function to a megasporophyll, but typically includes a stigma, and is fused, with ovules enclosed in the enlarged lower portion, the ovary.

In some basal angiosperm lineages, Degeneriaceae and Winteraceae, a carpel begins as a shallow cup where the ovules develop with laminar placentation, on the upper surface of the carpel. The carpel eventually forms a folded, leaf-like structure, not fully sealed at its margins. No style exists, but a broad stigmatic crest along the margin allows pollen tubes access along the surface and between hairs at the margins.

Two kinds of fusion have been distinguished: postgenital fusion that can be observed during the development of flowers, and congenital fusion that cannot be observed i. e., fusions that occurred during phylogeny. But it is very difficult to distinguish fusion and non-fusion processes in the evolution of flowering plants. Some processes that have been considered congenital (phylogenetic) fusions appear to be non-fusion processes such as, for example, the de novo formation of intercalary growth in a ring zone at or below the base of primordia. Therefore, "it is now increasingly acknowledged that the term 'fusion, ' as applied to phylogeny (as in 'congenital fusion') is ill-advised. "

Gynoecium Position

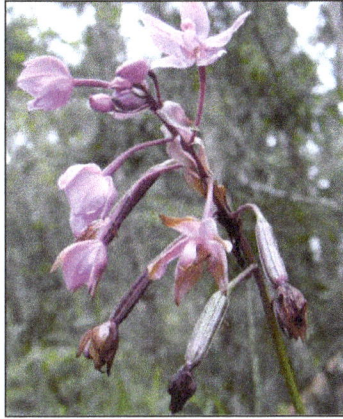

Flowers and fruit (capsules) of the ground orchid,
Spathoglottis plicata, illustrating an inferior ovary.

Basal angiosperm groups tend to have carpels arranged spirally around a conical or dome-shaped receptacle. In later lineages, carpels tend to be in whorls.

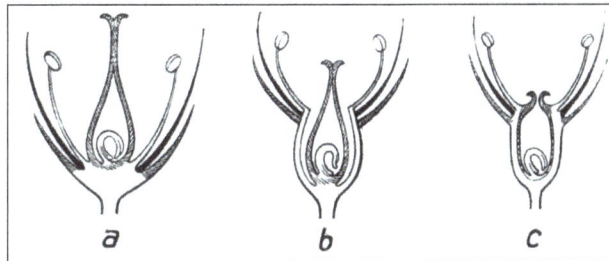

Illustration showing longitudinal sections through
hypogynous (a), perigynous (b), and epigynous (c) flowers.

The relationship of the other flower parts to the gynoecium can be an important systematic and taxonomic character. In some flowers, the stamens, petals, and sepals are often said to be "fused" into a "floral tube" or hypanthium. However, as Leins & Erbar (2010) pointed out, "the classical view that the wall of the inferior ovary results from the "congenital" fusion of dorsal carpel flanks and the floral axis does not correspond to the ontogenetic processes that can actually be observed. All that can be seen is an intercalary growth in a broad circular zone that changes the shape of the floral axis (receptacle). " And what happened during evolution is not a phylogenetic fusion but the formation of a unitary intercalary meristem. Evolutionary developmental biology investigates such developmental processes that arise or change during evolution.

If the hypanthium is absent, the flower is hypogynous, and the stamens, petals, and sepals are all attached to the receptacle below the gynoecium. Hypogynous flowers are often referred to as having a superior ovary. This is the typical arrangement in most flowers.

If the hypanthium is present up to the base of the style(s), the flower is epigynous. In an epigynous flower, the stamens, petals, and sepals are attached to the hypanthium at the top of the ovary or, occasionally, the hypanthium may extend beyond the top of the ovary. Epigynous flowers are often referred to as having an inferior ovary. Plant families with epigynous flowers include orchids, asters, and evening primroses.

Between these two extremes are perigynous flowers, in which a hypanthium is present, but is either free from the gynoecium (in which case it may appear to be a cup or tube surrounding the gynoecium) or connected partly to the gynoecium (with the stamens, petals, and sepals attached to the hypanthium part of the way up the ovary). Perigynous flowers are often referred to as having a half-inferior ovary (or, sometimes, partially inferior or half-superior). This arrangement is particularly frequent in the rose family and saxifrages.

Occasionally, the gynoecium is born on a stalk, called the gynophore, as in *Isomeris arborea.*

Placentation

Within the ovary, each ovule is born by a placenta or arises as a continuation of the floral apex. The placentas often occur in distinct lines called lines of placentation. In monocarpous or apocarpous gynoecia, there is typically a single line of placentation in each ovary. In syncarpous gynoecia, the lines of placentation can be regularly spaced along the wall of the ovary (parietal placentation), or near the center of the ovary. In the latter case, separate terms are used depending on whether or not the ovary is divided into separate locules. If the ovary is divided, with the ovules born on a line of placentation at the inner angle of each locule, this is axile placentation. An ovary with free central placentation, on the other hand, consists of a single compartment without septae and the ovules are attached to a central column that arises directly from the floral apex (axis). In some cases a single ovule is attached to the bottom or top of the locule (basal or apical placentation, respectively).

The Ovule

Longitudinal section of carpellate flower of squash showing
ovary, ovules, stigma, style, and petals.

In flowering plants, the ovule is a complex structure born inside ovaries. The ovule initially consists of a stalked, integumented megasporangium (also called the nucellus). Typically, one cell in the megasporangium undergoes meiosis resulting in one to four megaspores. These develop into a megagametophyte (often called the embryo sac) within the ovule. The megagametophyte typically develops a small number of cells, including two special cells, an egg cell and a binucleate central cell, which are the gametes involved in double fertilization. The central cell, once fertilized by a sperm cell from the pollen becomes the first cell of the endosperm, and the egg cell once fertilized become the zygote that develops into the embryo. The gap in the integuments through which the pollen tube enters to deliver sperm to the egg is called the micropyle. The stalk attaching the ovule to the placenta is called the funiculus.

Role of the Stigma and Style

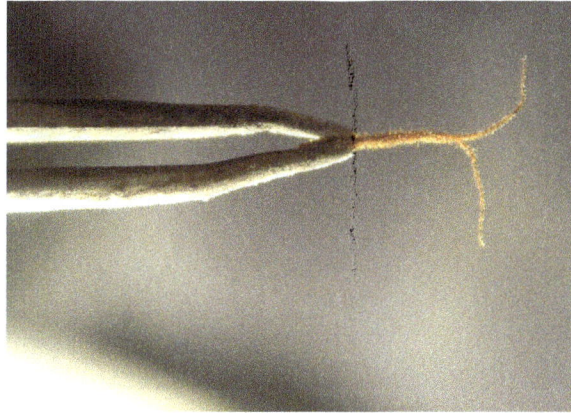

Stigmas and style of Cannabis sativa held in a pair of forceps.

Stigmas can vary from long and slender to globe-shaped to feathery. The stigma is the receptive tip of the carpel, which receives pollen at pollination and on which the pollen grain germinates. The stigma is adapted to catch and trap pollen, either by combining pollen of visiting insects or by various hairs, flaps, or sculpturings.

Stigma of a Crocus flower.

The style and stigma of the flower are involved in most types of self incompatibility reactions. Self-incompatibility, if present, prevents fertilization by pollen from the same plant or from genetically similar plants, and ensures outcrossing.

Ovary

In the flowering plants, an ovary is a part of the female reproductive organ of the flower or gynoecium. Specifically, it is the part of the pistil which holds the ovule(s) and is located above or below or at the point of connection with the base of the petals and sepals. The pistil may be made up of one carpel or of several fused carpels (e. g. dicarpel or tricarpel), and therefore the ovary can contain part of one carpel or parts of several fused carpels. Above the ovary is the style and the stigma, which is where the pollen lands and germinates to grow down through the style to the ovary, and, for each individual pollen grain, to fertilize one individual ovule. Some wind pollinated flowers have much reduced and modified ovaries.

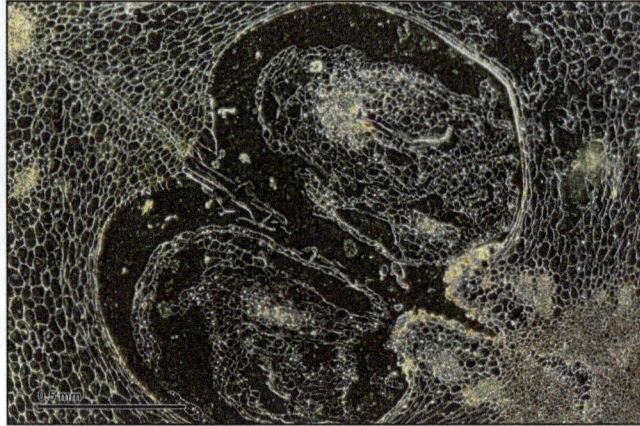

Cross section of *Tulip* ovary.

Fruits

A fruit is the ripened ovary or ovaries—together with seeds—from one or more flowers. The fruits of a plant are responsible for dispersing the seeds that contain the embryo and protecting the seeds as well. In many species, the fruit incorporates some surrounding tissues, or is dispersed with some non-fruit tissues.

Parts of the Ovary

Locules are chambers within the ovary of the flower and fruits. The locules contain the ovules (seeds), and may or may not be filled with fruit flesh. Depending on the number of locules in the ovary, fruits can be classified as uni-locular (unilocular), bi-locular, tri-locular or multi-locular. Some plants have septa between the carpels; the number of locules present in a gynoecium may be equal to or less than the number of carpels, depending on whether septa are present.

The syncarpous ovary of this melon is made up of four carpels, and has one locule.

In this Peganum harmala, the ovary of a fruit has split into valves.

The ovules are attached to parts of the interior ovary walls called the placentae. Placental areas occur in various positions, corresponding to various parts of the carpels that make up the ovary.

An obturator is present in the ovary of some plants, near the micropyle of each ovule. It is an outgrowth of the placenta, important in nourishing and guiding pollen tubes to the micropyle.

The seeds in a tomato fruit grow from placental areas at the interior of the ovary. (This is axile placentation in a bi-locular fruit.)

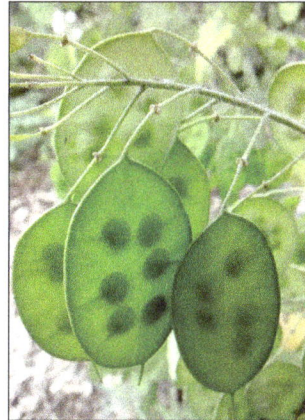

The placentae in *Lunaria* are along the margins of the fruit, where two carpels fuse. (This is parietal placentation in a bi-locular fruit.)

The ovary of some types of fruit is dehiscent; the ovary wall splits into sections called valves. There is no standard correspondence between the valves and the position of the septa; the valves may separate by splitting the septa (septicidal dehiscence), or by spitting between them (loculicidal dehiscence), or the ovary may open in other ways, as through pores or because a cap falls off.

The valves of *Lunaria* fruit fall to reveal a septum that was between the two carpels of the ovary.

Classification based on Position

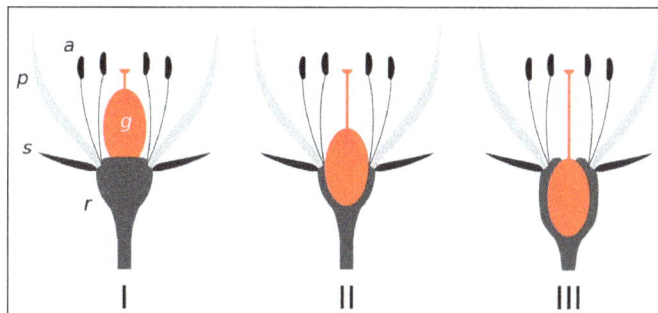

Ovary insertion: I superior II half-inferior III inferior.
a androecium *g* gynoecium *p* petals *s* sepals *r* receptacle.
The insertion point is where *a*, *p*, and *s* converge.

The terminology of the positions of ovaries is determined by the insertion point, where the other floral parts (perianth and androecium) come together and attach to the surface of the ovary. If the ovary is situated above the insertion point, it is superior; if below, inferior.

Superior Ovary

A superior ovary is an ovary attached to the receptacle above the attachment of other floral parts. A superior ovary is found in types of fleshy fruits such as true berries, drupes, etc. A flower with this arrangement is described as hypogynous. Examples of this ovary type include the legumes (beans and peas and their relatives).

Half-inferior Ovary

A half-inferior ovary (also known as "half-superior", "subinferior, " or "partially inferior, ") is embedded or surrounded by the receptacle. This occurs in flowers of the Lythraceae family, which includes the Crape Myrtles. Such flowers are termed perigynous or half-epigynous. In some classifications, half-inferior ovaries are not recognized and are instead grouped with either the superior or inferior ovaries.

More specifically, a half-inferior ovary has nearly equal portions of ovary above and below the insertion point. Other varying degrees of inferiority can be described by other fractions. For instance, a "one-fifth inferior ovary" has approximately one fifth of its length under the insertion point. Likewise, only one quarter portion of a "three-quarters inferior ovary" is above the insertion.

Inferior Ovary

An inferior ovary lies below the attachment of other floral parts. A pome is a type of fleshy fruit that is often cited as an example, but close inspection of some pomes (such as *Pyracantha*) will show that it is really a half-inferior ovary. Flowers with inferior ovaries are termed epigynous. Some examples of flowers with an inferior ovary are orchids (inferior capsule), *Fuchsia* (inferior berry), banana (inferior berry), Asteraceae (inferior achene-like fruit, called a cypsela) and the pepo of the squash, melon and gourd (Cucurbitaceae) family.

Stigma

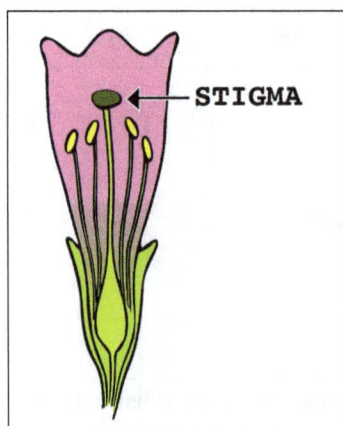

Diagram of stigma.

The stigma is the receptive tip of a carpel, or of several fused carpels, in the gynoecium of a flower.

Stigma of a *Tulipa* species, with pollen.

Closeup of stigma surrounded by stamen of White Lilium 'Stargazer' (the 'Stargazer lily').

The stigma, together with the style and ovary comprises the pistil, which in turn is part of the gynoecium or female reproductive organ of a plant. The stigma forms the distal portion of the style or stylodia. The stigma is composed of stigmatic papillae, the cells which are receptive to pollen. These may be restricted to the apex of the style or, especially in wind pollinated species, cover a wide surface.

The stigma receives pollen and it is on the stigma that the pollen grain germinates. Often sticky, the stigma is adapted in various ways to catch and trap pollen with various hairs, flaps, or sculpturings. The pollen may be captured from the air (wind-borne pollen, anemophily), from visiting insects or other animals (biotic pollination), or in rare cases from surrounding water (hydrophily). Stigma can vary from long and slender to globe shaped to feathery.

Pollen is typically highly desiccated when it leaves an anther. Stigma have been shown to assist in the rehydration of pollen and in promoting germination of the pollen tube. Stigma also ensure proper adhesion of the correct species of pollen. Stigma can play an active role in pollen discrimination and some self-incompatibility reactions, that reject pollen from the same or genetically similar plants, involve interaction between the stigma and the surface of the pollen grain.

Shape

The stigma is often split into lobes, e. g. trifid (three lobed), and may resemble the head of a pin (capitate), or come to a point (punctiform). The shape of the stigma may vary considerably:

Corn stigma called "silk".

Stigma Shapes

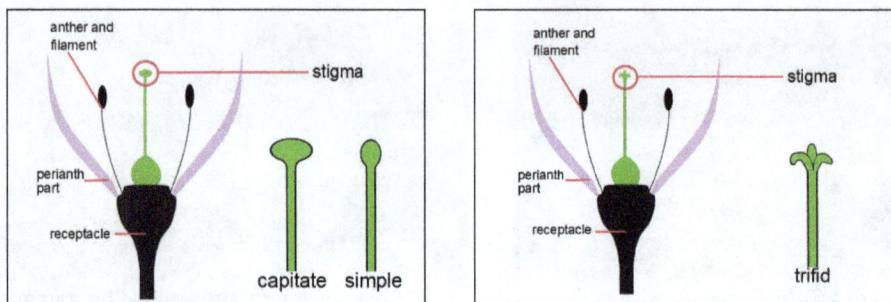

Capitate and simple.   Trifid.

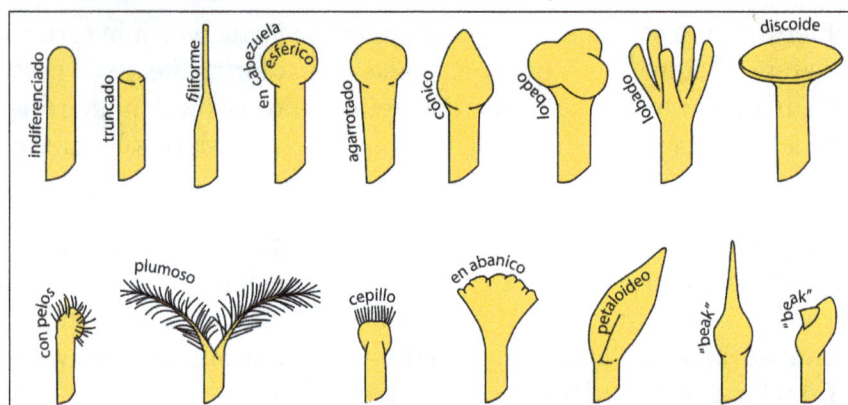

Style

Structure

The style is a narrow upward extension of the ovary, connecting it to the stigmatic papillae. It may be absent in some plants in the case the stigma is referred to as sessile. Styles are generally tube-like—either long or short. The style can be open (containing few or no cells in the central portion) with a central canal which may be filled with mucilage. Alternatively the style may be closed (densely packed with cells throughout). Most syncarpous monocots and some eudicots have open styles, while many syncarpous eudicots and grasses have closed (solid) styles containing specialised secretory transmitting tissue, linking the stigma to the centre of the ovary. This forms a nutrient rich tract for pollen tube growth.

Where there are more than one carpel to the pistil, each may have a separate style-like stylodium, or share a common style. In Irises and others in the family Iridaceae, the style divides into three petal-like (petaloid) style branches (sometimes also referred to as 'stylodia'), almost to the base of the style and is called tribrachiate. These are flaps of tissue, running from the perianth tube above the sepal. The stigma is a rim or edge on the underside of the branch, near the end lobes. Style branches also appear on Dietes, Pardanthopsis and most species of Moraea.

In Crocuses, there are three divided style branches, creating a tube. Hesperantha has a spreading style branch. Alternatively the style may be lobed rather than branched. Gladiolus has a bi-lobed style branch (bilobate). Freesia, Lapeirousia, Romulea, Savannosiphon and Watsonia have bifuracated (two branched) and recurved style branches.

Style Morphology

Iris versicolor showing three structures with two overlapping lips,
an upper petaloid style branch and a lower tepal, enclosing a stamen.

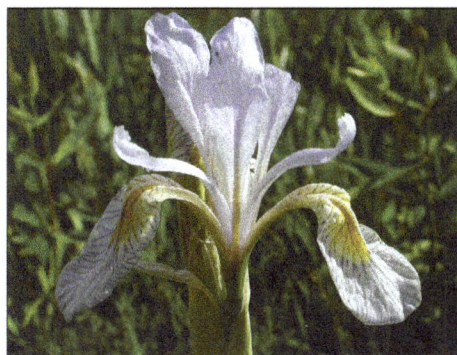

Iris missouriensis showing
the pale blue style branch
above the drooping petal.

The feathery stigma of
Crocus speciosus has branches
corresponding to three carpels.

Attachment to the Ovary

Style Position

Terminal (apical).

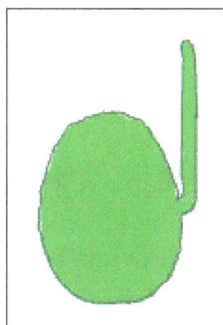

Lateral.

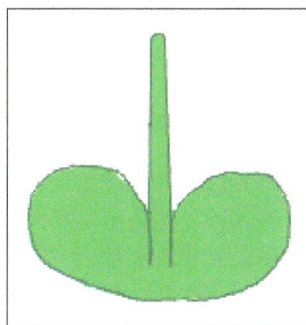

Gynobasic.

May be terminal (apical), subapical, lateral, gynobasic, or subgynobasic. Terminal (apical) style position refers to attachment at the apex of the ovary and is the commonest pattern. In the subapical pattern the style arises to the side slightly below the apex. a lateral style arises from the side of the ovary and is found in Rosaceae. The gynobasic style arises from the base of the ovary, or between the ovary lobes and is characteristic of Boraginaceae. Subgynobasic styles characterise Allium.

Pollination

Pollen tubes grow the length of the style to reach the ovules, and in some cases self-incompatibility reactions in the style prevent full growth of the pollen tubes. In some species, including *Gasteria* at least, the pollen tube is directed to the micropyle of the ovule by the style.

Male Reproductive Parts of Flower

Collectively, the male parts of the flower are called the stamen. Individually, the male reproductive parts are called the anther and the filament. The filament, which resembles a hair, holds a round pouch on top of it called the anther. The anther produces pollen, which is held in the small round pouches that sit on top of the filament.

Stamen

The stamen is the pollen-producing reproductive organ of a flower. Collectively the stamens form the androecium.

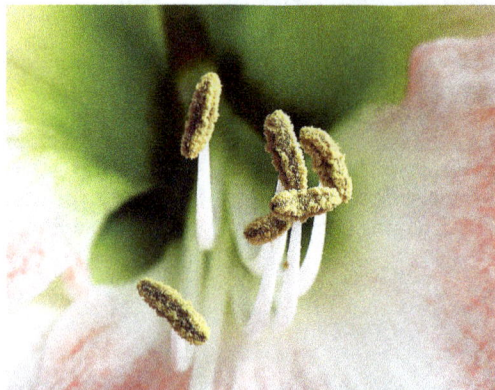

Stamens of a Hippeastrum with white filaments
and prominent anthers carrying pollen.

Morphology and Terminology

A stamen typically consists of a stalk called the filament and an anther which contains *microsporangia*. Most commonly anthers are two-lobed and are attached to the filament either at the base or in the middle area of the anther. The sterile tissue between the lobes is called the connective, an extension of the filament containing conducting strands. A pollen grain develops from a microspore in the microsporangium and contains the male gametophyte.

The stamens in a flower are collectively called the androecium. The androecium can consist of as few as one-half stamen (i. e. a single locule) as in Canna species or as many as 3, 482 stamens which have been counted in Carnegiea gigantea. The androecium in various species of plants forms a great variety of patterns, some of them highly complex. It generally surrounds the gynoecium and is surrounded by the perianth. A few members of the family Triuridaceae, particularly Lacandonia schismatica, are exceptional in that their gynoecia surround their androecia.

Variation in Morphology

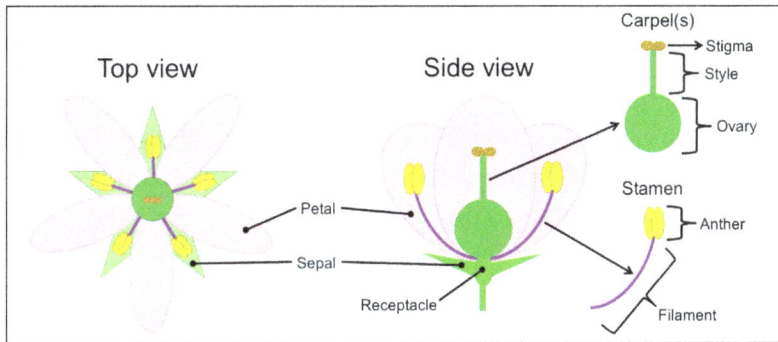

Stamens, with distal anther attached to the filament stalk, in context of floral anatomy.

Depending on the species of plant, some or all of the stamens in a flower may be attached to the petals or to the floral axis. They also may be free-standing or fused to one another in many different ways, including fusion of some but not all stamens. The filaments may be fused and the anthers free, or the filaments free and the anthers fused. Rather than there being two locules, one locule of a stamen may fail to develop, or alternatively the two locules may merge late in development to give a single locule. Extreme cases of stamen fusion occur in some species of *Cyclanthera* in the family Cucurbitaceae and in section *Cyclanthera* of genus *Phyllanthus* (family Euphorbiaceae) where the stamens form a ring around the gynoecium, with a single locule.

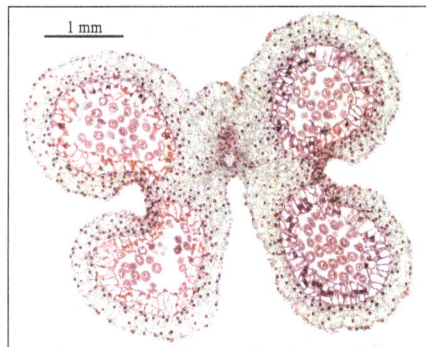

Cross section of a *Lilium* stamen, with four
locules surrounded by the tapetum.

Pollen Production

A typical anther contains four microsporangia. The *microsporangia* form sacs or pockets (*locules*) in the anther (anther sacs or pollen sacs). The two separate locules on each side of an anther may fuse into a single locule. Each microsporangium is lined with a nutritive tissue layer called the *tapetum* and initially contains diploid pollen mother cells. These undergo meiosis to form haploid spores. The spores may remain attached to each other in a tetrad or separate after meiosis. Each microspore then divides mitotically to form an immature microgametophyte called a pollen grain.

The pollen is eventually released when the anther forms openings (dehisces). These may consist of longitudinal slits, pores, as in the heath family (Ericaceae), or by valves, as in the barberry family (Berberidaceae). In some plants, notably members of Orchidaceae and Asclepiadoideae, the pollen remains in masses called pollinia, which are adapted to attach to particular pollinating agents such as birds or insects. More commonly, mature pollen grains separate and are dispensed by wind or water, pollinating insects, birds or other pollination vectors.

Pollen of angiosperms must be transported to the stigma, the receptive surface of the *carpel*, of a compatible flower, for successful pollination to occur. After arriving, the pollen grain (an immature microgametophyte) typically completes its development. It may grow a pollen tube and undergoing mitosis to produce two sperm nuclei.

Sexual Reproduction in Plants

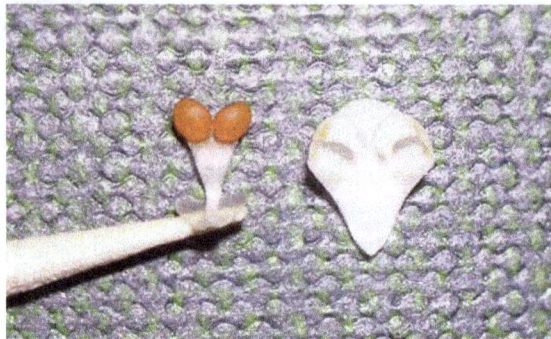

Stamen with pollinia and its anther cap. *Phalaenopsis* orchid.

In the typical flower (that is, in the majority of flowering plant species) each flower has both carpels and stamens. In some species, however, the flowers are unisexual with only carpels or stamens (monoecious = both types of flowers found on the same plant; dioecious = the two types of flower found only on different plants). A flower with only stamens is called androecious. A flower with only carpels is called gynoecious.

A flower having only functional stamens and lacking functional carpels is called a staminate flower, or (inaccurately) male. A plant with only functional carpels is called pistillate, or (inaccurately) female.

An abortive or rudimentary stamen is called a staminodium or staminode, such as in Scrophularia nodosa.

The carpels and stamens of orchids are fused into a column. The top part of the column is formed by the anther, which is covered by an anther cap.

Descriptive Terms

Stamen

Stamens can also be adnate (fused or joined from more than one whorl):

- Epipetalous: Adnate to the corolla.

- Epiphyllous: Adnate to undifferentiated tepals (as in many Liliaceae).

They can have different lengths from each other:

- Didymous: Two equal pairs.

- didynamous: Occurring in two pairs, a long pair and a shorter pair.

- Tetradynamous: Occurring as a set of six stamens with four long and two shorter ones.

Respective to the rest of the flower (perianth):

- Exserted: Extending beyond the corolla.

- Included: Not extending beyond the corolla.

They may be arranged in one of two different patterns:

- Spiral.

- Whorled: One or more discrete whorls (series).

They may be arranged, with respect to the petals:

- Diplostemonous: In two whorls, the outer alternating with the petals, while the inner is opposite the petals.

- Haplostemenous: Having a single series of stamens, equal in number to the proper number of petals and alternating with them.

- Obdiplostemonous: In two whorls, with twice the number of stamens as petals, the outer opposite the petals, inner opposite the sepals, e. g. Simaroubaceae.

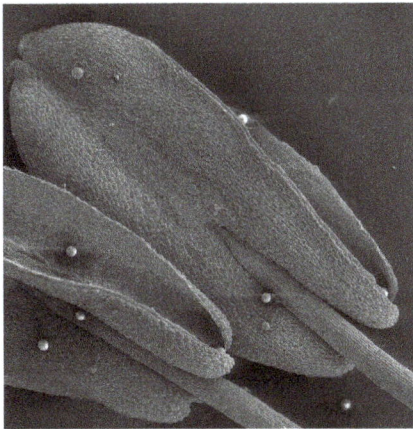

Scanning electron microscope image of *Pentas lanceolata* anthers, with pollen grains on surface.

Lily stamens with prominent red anthers and white filaments.

Connective

Where the connective is very small, or imperceptible, the anther lobes are close together, and the connective is referred to as discrete, e. g. Euphorbia pp., Adhatoda zeylanica. Where the connective separates the anther lobes, it is called divaricate, e. g. Tilia, Justicia gendarussa. The connective may also be a long and stalk-like, crosswise on the filament, this is a distractile connective, e. g. Salvia. The connective may also bear appendages, and is called appendiculate, e. g. Nerium odorum and some other species of Apocynaceae. In Nerium, the appendages are united as a staminal corona.

Filament

A column formed from the fusion of multiple filaments is known as an androphore. Stamens can be connate (fused or joined in the same whorl) as follows:

- Extrorse: Anther dehiscence directed away from the centre of the flower. Cf. introrse, directed inwards, and latrorse towards the side.

- Monadelphous: Fused into a single, compound structure.

- Declinate: Curving downwards, then up at the tip (also – declinate-descending)

- Diadelphous: Joined partially into two androecial structures.

- Pentadelphous: Joined partially into five androecial structures.

- Synandrous: Only the anthers are connate (such as in the Asteraceae). The fused stamens are referred to as a synandrium.

Anther

Anther shapes are variously described by terms such as linear, rounded, sagittate, sinuous, or reniform.

The anther can be attached to the filament's connective in two ways:

- Basifixed: Attached at its base to the filament

    ◦ Pseudobasifixed: A somewhat misnomer configuration where connective tissue extends in a tube around the filament tip

- Dorsifixed: Attached at its center to the filament, usually versatile (able to move)

Flower Development

The flower is the most complex structure of plants. Flowers distinguish the most recently diverged plant lineage, the angiosperms or flowering plants, from the other land plants. Embryophytes originated approximately 450 million years before present (MYBP) and have as distinctive features a thick cuticle resistant to desiccation, sporopollenin, pores or true stomata that aid in gas exchange, a glycolate oxidase system that improves carbon fixation at high oxygen tensions, and importantly, distinctive multicellular diploid (sporophytic) and haploid (gametophytic) stages within their life cycles. The major extant land plant lineages are Bryophytes (Liverworts, Hornworts and Mosses), which do not have a vascular system, and Tracheophytes, vascular plants. Within the large latter group, Lycophytes, ferns, and seed bearing plants (Spermatophytes) can be distinguished. The Spermatophyte group has been further divided into Gymnosperms (originating 380–325 MYBP) and Angiosperms. According to the fossil record, flower-like structures originated 160–147 MYBP. A general trend within land plant evolution is the appearance of heterospory: the existence of a megagametophyte, including the female gametes, and a microgametophyte, including the male gametes, a progressive reduction in gametophyte size (sexual reproductive structures), and within the seed plants, the presence of a diploid embryo. While these characteristics are shared among both extant and extinct seed plant lineages, the defining features of the angiosperm flower are: (1) a closed carpel bearing the ovules, which are each generally comprised of two integuments and (2) a nucellus that contains the embryo sac within which, after double fertilization, a diploid embryo and a triploid endosperm (nutritional tissue for the embryo) will develop to form a seed. Another characteristic of angiosperms is true hermaphroditism.

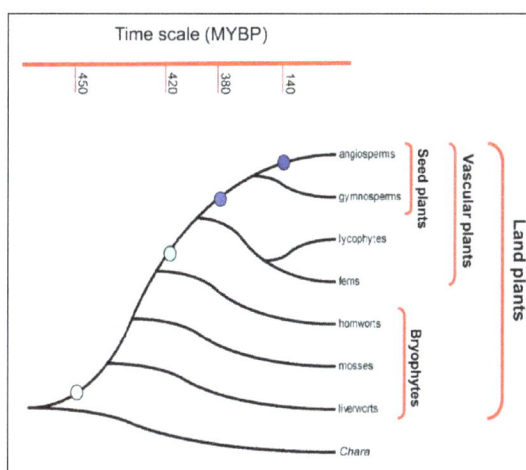

Phylogenetic context of Arabidopsis thaliana: Evolutionary history of land plants.

Phylogenetic tree of land plant evolution with some speciation events shown as colored nodes. White node, origin of land plants; light blue node, origin of vascular plants; blue node, origin of seed plants; dark blue node, origin of flowering plants. Here, *Chara* spp. from the green algae order Charales is the outgroup, since it has been used to root several recent molecular land plant phylogenies. The topology of this tree is based on studies by Soltis et al. and Nickrent et al.

Flower structure has been studied in a variety of ways. Studies of the natural history and evolutionary biology of flowers have emphasized understanding the ultimate (evolutionary) causes of the wide range of variants such as color, symmetry, meristic arrangements (e. g. flower organ number), size, pollination syndrome, etc. Other studies have addressed the cellular, tissue type, morphological and physical factors that can account for both the phenotypic plasticity and developmental constraints in flower form. A different approach flourished in the late 1980s and early 1990s, the molecular genetics of flower development in two model eudicot species: Arabidopsis thaliana and Antirrhinum majus.

Genetic studies of floral homeotic mutants in both plant species yielded the now classic combinatorial ABC developmental model for floral organ determination. While much work has been and continues to be done in Antirrhinum and other eudicot species, including Petunia hybrida, the genomic and life-cycle characteristics of Arabidopsis make it the preferred experimental system for in-depth studies on the molecular components underlying cell differentiation and morphogenesis during flower development.

The basic floral architecture is mostly conserved among the so-called core eudicots, that make up over 73% of extant flowering plants including Arabidopsis. Flowers within this group generally have four concentric whorls of organs that are specified, from the outside to the center of the flower, in the sequence: sepals, petals, stamens, and carpels. Arabidopsis has this typical floral architecture. An interesting exception to the conserved floral ground plan of eudicots is found in a Mexican rainforest monocotyledon, Lacandonia schismatica (Triuridaceae), which bears central stamens surrounded by carpels.

Even though the basic floral architecture is overall conserved among core eudicots, variation in the symmetry and size of flowers, the number of whorls of each organ type, the number of organs per whorl, and their arrangement, size, shape and color is common.

The overall conservation of the flower plan suggests that robust gene regulatory network (GRN) modules controlling the basic features of flower development were established early in the evolution of angiosperms and have persisted in the great majority of lineages throughout 140 million years of flower evolution. Recent integrated approaches to study the concerted action of the molecular components in flower development, have led to a hypothesis that helps explain such robustness and conservation at the level of the GRN underlying floral organ specification. However structural (e. g., mechanical) constraints could also be involved in conserving floral architecture. Approaches that integrate genetic and structural aspects of flowers should be pursued further to understand flower development in Arabidopsis and other angiosperms.

Structural Organization of the Inflorescence Meristem

During the vegetative phase of the Arabidopsis life cycle, the shoot apical meristem (SAM) produces leaves on its flanks and on transition to flowering, the shoot bolts and the SAM becomes the inflorescence shoot apical meristem (IM). On bolting, some of the pre-existing leaf primordia become cauline leaves subtending lateral inflorescence shoots (paraclades) and the shoot apex starts to produce flowers. A primary IM produces lateral meristems that may go on to produce flowers or secondary inflorescences. Arabidopsis inflorescences are subtended by fully developed bracts, but flowers only by rudimentary ones. It is generally said that the IM generates the floral meristems (FM) on its flanks, but to be more precise, Arabidopsis FM are formed in the axils of the rudimentary bracts.

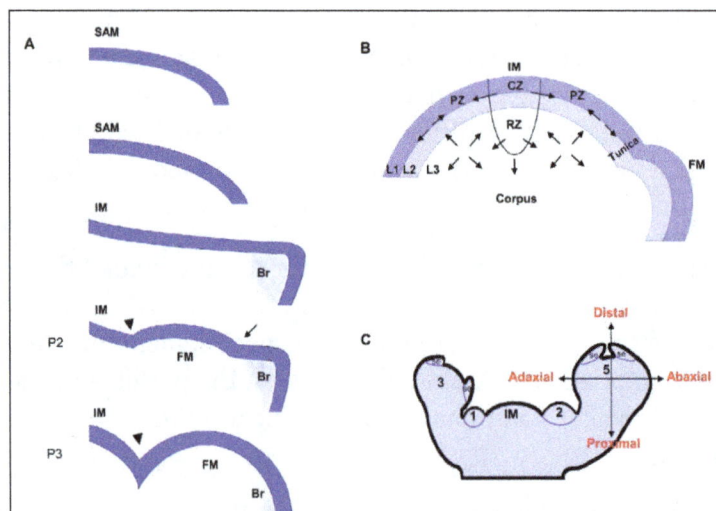

Schematic representation of the shoot apical meristem (SAM): The inflorescence shoot apical meristem and floral meristem.

(A) Diagram outlining the geometry of the inflorescence shoot apical meristem (IM) and flower meristem (FM) during the first stages of development of the latter. On the flank of the IM a first bulge that corresponds to the rudimentary bract (Br) appears. In its axil, a second bulge forms and this continues to grow engulfing the first one and forming the FM proper. Theses stages of FM development correspond to P2 and P3 according to Reddy et al. The arrow and arrowhead indicate the first and second visible grooves respectively.

(B) Three distinctive zones make up the IM: the central zone (CZ) which contains the stem cells; the peripheral zone (PZ) on the flanks of the CZ that gives rise to the bract and floral primordia; and the rib zone (RZ) underneath the CZ that yields stem tissue. Three cell layers are distinguished: L1 and L2 layers constitute the tunica and include portions of both the CZ and the PZ. The rest of the cells form the L3 layer or corpus. In L1 and L2, cell divisions are anticlinal, while in L3 they occur in all directions (arrows, direction of cell division). The structure is maintained in the FM.

(C) Schematic representation of the boundary zones (blue lines) and axes of polarity during floral development with the differentiation of sepals (se) from the floral primordium illustrated.

The SAM of the Arabidopsis inflorescence consists of a small dome of cells organized into different regions with different gene expression profiles, cellular behaviors and structures. The tunica at the SAM surface and corpus are distinguished on the basis of cell division planes. In Arabidopsis, the tunica consists of two clonally distinct cell layers called L1 and L2. Cell divisions within these meristem layers are exclusively anticlinal and the new cell walls are formed perpendicular to the surface of the meristem. The progeny of cells in the L1 will therefore remain in this same layer within the meristem similar to the underlying L2 progeny. Since outside the meristem the L1 derived cells continue to divide only anticlinally the L1 eventually gives rise to epidermal cells. The cells originating from L2 also divide periclinally (outside the SAM) and contribute for example to the leaf mesophyll or stem ground tissue formation during organogenesis. This is also the germ line in the angiosperm SAM. Below the tunica, cell divisions are both anticlinal and periclinal. This region of the SAM is the corpus or L3 from which the innermost tissues, like vascular tissues, are formed.

The SAM is also organized into three different cytohistological zones each with characteristic cytoplasmic densities and cell division rates: the central zone (CZ), the peripheral zone (PZ) surrounding the CZ and the rib zone (RZ) underneath the CZ.

Flower primordia are derived from the PZ of the IM and are initiated from a block of four so-called founder cells. This estimate was based on sector boundary analysis. However, using a non-invasive replica method and a 3-D reconstruction algorithm, Kwiatkowska argues that more cells are assigned to the flower primordium, and this is consistent with the observations by Grandjean et al. The difference could be due to the fact that not all of the cells estimated to be involved in the latter approaches are incorporated into the flower meristem proper. Some of them may form a part of the subtending rudimentary bract.

The first cells produced by the RZ following the transition to flowering are rectangular with their long axis perpendicular to the major axis of the stem, but the subsequent elongation of these cells reverses this situation. The RZ gives rise to stem tissue. The CZ encompasses the reservoir of stem cells that divide less frequently than cells at the periphery. The CZ maintains itself and yields daughter cells that form both the PZ and RZ. Fifteen stages of Arabidopsis flower development have been distinguished. The first stages of flower meristem development are: stage 1, when a flower buttress arises, stage 2 when the flower meristem is formed and stage 3 when sepal primordia appear. Recently researchers have been able to study early flower meristem development in greater detail and have proposed subdividing stage 1.

Floral Organ Primordia

Once a flower primordium is initiated, the geometry changes and a rapid and coordinated burst of cell expansion and division occurs in three dimensions generating a concentric group of cells as an almost spherical flower primordium, from which all floral tissues are derived. Jenik and Irish found that the regulation of cell divisions during early and late stages of flower development seems to depend upon different mechanisms. Early in flower development, when the floral meristem of Arabidopsis is divided into four concentric rings (each with a characteristic multigenic expression profile), cell division patterns depend upon the cell's radial position in the floral meristem, and not on the future identity of the floral organ to be formed in each ring. After stage 6, during organogenesis, the ABC homeotic genes seem to control the rate and orientation of cell divisions. As a result, the continuity of the concentric rings is broken giving distinct floral organ primordia within each whorl, then cells subdifferentiate into distinct types within each organ. The initiation and identity of floral organs are also regulated by different and largely independent molecular modules. This is suggested, for example, by the fact that conversion of petals into sepal-like organs in mutant plants does not alter the number of cells involved in their initiation.

Tissues of floral organs are organized according to coordinated patterns and rates of cell division in the different cell layers of the meristem that dynamically acquire distinct fates. Clonal analysis shows that L1 contributes to the epidermis, the stigma, part of the transmitting tract and the integument of the ovules, while L2 and L3 contribute to the mesophyll and other internal tissues.

Sector boundary analysis of surface cells has shown that sepals and carpels are initiated from eight cells, stamens from four cells, and petals from two cells. Each organ primordium arises as a set of cells separated by boundary regions of slow-dividing cells. Flower development ends when mature organs are formed and all the flower meristem cells are used up.

Stages of Flower Development

STAGE 1: The first sign of flower primordium formation is the bulging of the peripheral surface of the IM in a lateral direction. This stage was referred to as P1 by Reddy et al. It is hypothesized that a lateral protrusion formed during bulging is a rudimentary bract. At this early stage, growth is fast and strongly anisotropic, with maximal growth in a meridional (i. e. radial when viewed from the top of the meristem) direction eventually leading to formation of a shallow crease, which corresponds to the first visible groove and to the P2 stage of flower development. This shallow crease corresponds to the axil of the putative rudimentary bract. Soon after the bract is formed, another bulge occurs in its axil in an upward direction. This second bulging corresponds to the formation of a flower primordium proper and to stage P3 according to Reddy et al. This stage corresponds to stage 2 according to Smyth et al. Hence, during early stages of flower development in Arabidopsis, two types of primordia (bract and flower primordium proper) and organ boundaries are observed. The first boundary is the adaxial boundary of the rudimentary bract, while the second is the boundary between the IM and the flower primordium proper. The expression patterns of several genes confirm the developmental stages distinguished here.

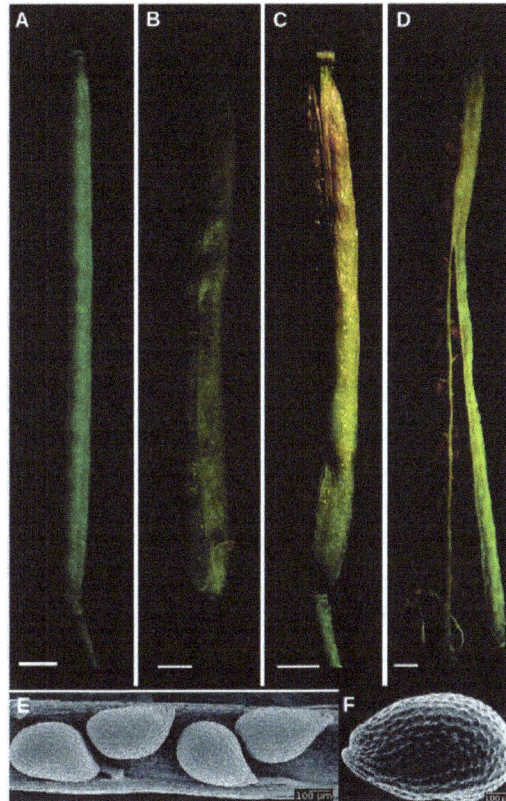

Stages 17 to 20 of Arabidopsis flower development.

In figure, (A) to (D) Photographs of developing and mature siliques at stages 17 (A), 18 (B), 19 (C), and 20 (D) of flower development. (E) SEM of seeds from a silique at stage 17. (F) Close-up view of a seed from a stage-20 dehiscent silique. All photographs are of Columbia-0 ecotype.

A significant increase in mitotic activity is observed upon formation of the primordium. The mitotic activity can be estimated as the increase in the number of cells per 24 h or the accompanying area growth rates on the condition that the mean cell size does not increase. During these early stages of flower development, periclinal cell divisions occur in the corpus while L1 and L2 cells only divide anticlinally. Hence, the two-layered tunica organization is maintained in the flower meristem, but all of its cells are mitotically active.

STAGE 2: During this stage, the hemispherical primordium continues to grow forming almost a right angle with the surface of the SAM, which itself lengthens and widens rebuilding the portion of the periphery that has been used for primordium formation. At this stage the flower primordium becomes clearly delimited from the IM, and starts to grow larger very quickly in all directions.

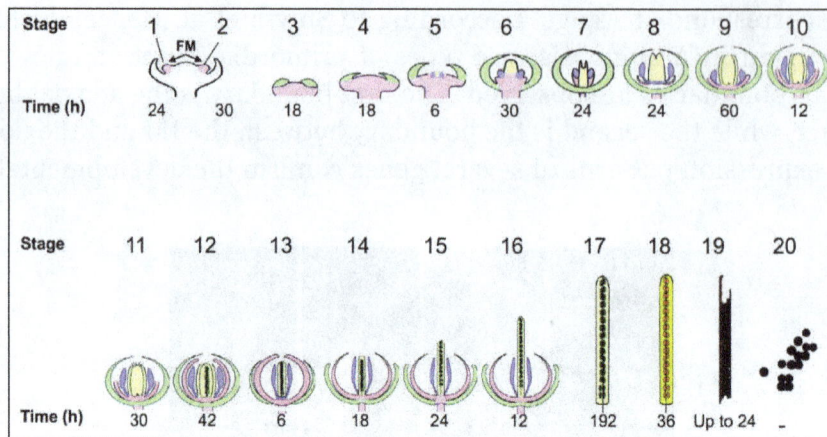

Summary of the 20 stages of flower development.

In above schematic representation of developmental stages of Arabidopsis flowers. Briefly, the flower primordium is formed at stages 1 and 2. At stage 3, sepal primordia are already visible and continue growing until they enclose the flower meristem (from stage 4 to 6). Meanwhile, at stage 5, petal and stamen primordia are beginning to be visible, and the gynoecium starts to form (stage 6). Organ development continues and by stage 9, stigmatic papillae arise at the top of the gynoecium. At stage 12, petals are similar in length to stamens. Anthesis occurs at stage 13, fertilization occurs, and the flower opens at stage 14. Siliques reach their maximum size and are green by stage 17, then they loose water and turn yellow (stage 18) until valves separate from dry siliques (stage 19) and seeds fall (stage 20). Floral meristems (FM), pink; sepals, green; petals, bright pink; stamens, blue; gynoecia, yellow; ovules, dark green; seeds orange and brown. Duration of each stage in hours (h) is given under the figures.

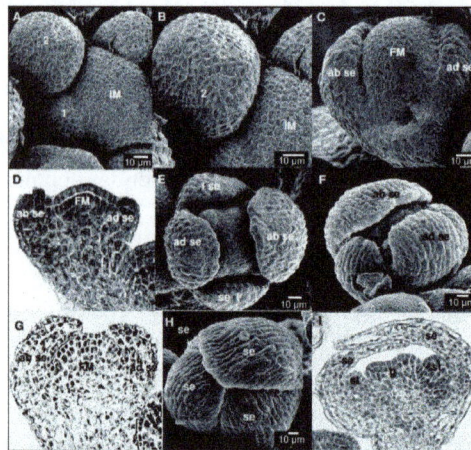

Stages 1 to 6 of Arabidopsis flower development.

(A) and (B) Inflorescence shoot apical meristem (IM) and floral meristem (FM) at stage 1 and 2 as indicated.

(C) and (D) Stage 3 FM showing abaxial (ab) and adaxial (ad) sepals (se).

(E) At stage 4, lateral sepals (I) shown growing perpendicularly to the abaxial and adaxial ones.

(F) and (Q) At stage 5, stamen primordia are visible (arrows) and sepals almost cover the rest of the meristem.

(H) Flower bud where sepals are covering the stamens and the gynoecium primordium.

(I) Section through a stage-6 flower primordium where the gynoecium (g), stamens (st), and sepals (se) are apparent.

Pictures are scanning electron micrographs (SEM), except (D), (Q) and (I) which are optical images of histological sections.

STAGE 3: This stage begins when sepal primordia become visible. By now the flower primordium is 30–35 μm in diameter and is becoming stalked with an incipient pedicel. It has also started to grow vertically. The two lateral (I) sepal primordia appear first, but are soon outgrown by the abaxial (ab) then the adaxial (ad) sepal primordia. Sepal primordia arise initially as ridges that lengthen and curve inwards until they begin to overtop the remaining dome-shaped portion of the flower primordium).

STAGE 4: During this stage, the elongation of the pedicel continues concurrently with an increase in the diameter of the developing flower primordium to 65–70 μm. The medial sepal primordia have already partly overtopped the remaining floral meristem.

STAGE 5: This stage is when the petal and stamen primordia become visible. Primordia of the four medial (long) stamens are first seen as wide outgrowths on the flanks of the central dome of the FM. The four petal primordia that arise between the sepals close to their base are just visible during this stage. The two lateral (short) stamens develop from primordia that appear later during this stage.

STAGE 6: The sepals grow to completely cover the floral bud and the primordia of the four long stamens bulge out and become distinct from the central dome of cells that comprise the FM. The two lateral stamen primordia arise slightly lower on the dome and develop later. The petal primordia grow somewhat but are still relatively small. A rim around the central dome of the flower primordium now begins to grow upward to produce an oval tube that will become the gynoecium.

STAGE 7: This stage begins when the growing primordia of the long stamens become stalked at their base. The stalks give rise to the filaments, and the wider upper region to the anthers. By this stage, petal primordia have become hemispherical although they are still relatively small (ca. 25 μm in diameter.

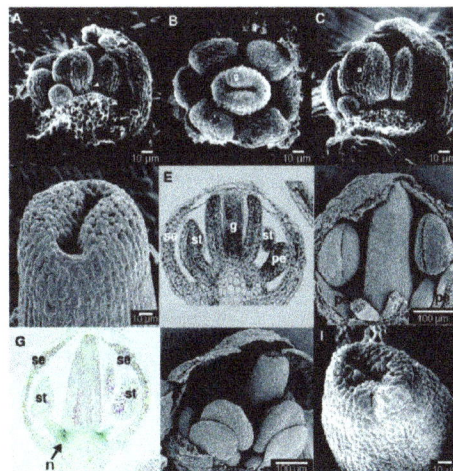

Stages 7 to 10 of Arabidopsis flower development.

(A) Stage 7 in which petal (pe) and stamen (arrowhead) primordia are indicated.

(B) Vertical view of the gynoecium (g) in a stage 7 floral primordium.

(C) to (E) Carpels and stamens at stage 8 of floral development are shown. Filament (f) and anther (a) regions of the stamen are differentiated (C) and a slot is formed at the tip of the style in the gynoecium (D). Section through the floral bud with sepals (se), petals (pe), stamens (st) and gynoecium (g) indicated (E).

(F) and (G) Floral bud at stage 9 in which petal primordia (pe) are indicated (F). Section through flower primordium (Q) in which *XALV. GUS* is shown staining nectaries (n).

(H) and (I) Stage 10 flowers. Flower bud showing the enlarged sepals which cover other floral organs, stalked petals and stamens, and developing carpels in the center (H). Stigma starts to be formed at the top of the gynoecium (I, arrows)

Bars = 10 μm except in (F) and (H). Images (A), (B), (C), (G) and (I) are of Lansberg *erecta* ecotype, from Smyth et al. provided by Dr J. Bowman. Some sepals were removed from flower buds shown in (A), (B), (C), (D), (F), (H) and (I). All images except (E) and (Q) are SEM. (D), (E), (F), (H) are of Columbia-o ecotype.

STAGE 8: The beginning of stage 8 is defined by another landmark in stamen development: anther locules are visible as convex protrusions on the inner (adaxial) surface of the long stamens. At this stage stamens are 55–60 μm long most of which is the developing anther. Locules also appear soon after in the short stamens. Petal growth now accelerates and petal primordia become apparent.

STAGE 9: This stage begins when the petal primordia elongate. There is a rapid lengthening of all organs especially of petals that acquire a tongue-like shape and increase in length from about 45 μm to up to 200 μm. Nectary glands appear and the stamens grow rapidly. By the end of stage 9, the medial stamens are around 300 μm long. Most of this growth occurs in the anther region, which still accounts for over 80% of their total length. At this stage the floral bud remains completely closed.

STAGE 10: The rapidly growing petals reach the top of the lateral stamens. The cap of papillae that will constitute the stigma starts to form at the top of the gynoecium.

STAGE 11: This stage begins when the upper surface of the gynoecium develops stigmatic papillae although their outward growth is limited at first to regions not in contact with the overlapping sepals. By the end of this stage petal primordia reach the top of the medial stamens.

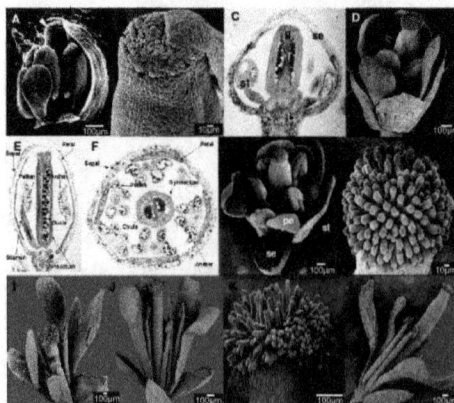

Stages 11 to 16 of Arabidopsis flower development.

(A) to (C) Stage 11 of flower development where the gynoecium develops stigmatic papillae (arrows) (A) and (B). Longitudinal section where sepals (se), stamen (st), and gynoecium (g) are indicated (C).

(D) to (F) Flower primordium at stage 12. Longitudinal (E) and transverse (F) sections showing all the organs as well as ovules and pollen grains.

(G) and (H) Flower anthesis at early stage 13 when the stigma (arrowhead) is already receptive (Q); a close-up view of the stigma (H).

(I) to (L) Flower primordium at stages 14 (I) and 15 where the gynoecium has begun to enlarge to form the silique (J). Close-up of a stage-15 stigma (K) and stage-16 flowers where sepals and petals are beginning to wither (L).

Bars = 100 μm. All images except (C), (E) and (F) are SEM. Images are of Columbia-0 ecotype, except (A) that is of Landsberg *erecta*.

STAGE 12: Petals continue to lengthen relatively rapidly. Lateral sepals continue to grow while the stamens and gynoecium lengthen coordinately. The anthers have almost reached their mature length of 350–400 μm and the filaments now lengthen rapidly. The upper part of the gynoecium differentiates into the style and a sharp boundary separates it from the cap of stigmatic papillae. Stage 12 ends when the sepals open.

STAGE 13: Petals become visible between the sepals and continue to elongate rapidly. The stigma is receptive at this stage. Stamen filaments extend even faster sothe stamens outstrip the gynoecium in length and self pollination takes place. The gynoecium is now mature and its three distinct regions can be distinguished: an apical stigma, a style, and a basal ovary. After pollination, pollen tubes grow to fertilize the ovules, the stamens extend above the stigma, and furrows at both valve/replum boundaries appear.

STAGE 14: This is also defined as the stage zero hours after flowering (0 HAF), and it marks the beginning of silique (the fertilized pistil or fruit) and seed development. Cells in the exocarp continue to divide anticlinally and expand longitudinally in the replum and the valve, where there is also some expansion in other directions. There is also division and expansion in the mesocarp and many chloroplasts develop.

STAGE 15: The stigma extends above the long anthers. In the carpel walls, cell division and expansion continue. The medial vascular bundles continue to grow and xylem lignifies, while the lateral bundles branch out through the mesocarp.

STAGE 16: At this stage the silique is twice as long as a stage-13 pistil. Petals and sepals wither and tissues in the silique continue expanding.

STAGE 17: This stage is defined by the abscission of the senescent floral organs from the silique, ~2 days after fertilization. The green silique grows to reach its final length and matures, a phase lasting about 8 days making this the longest stage. The dehiscence zone also differentiates.

STAGE 18: The silique begins to yellow from the tip to the base. One of the endocarp cell layers (the second from the inside) lignifies further, and the inner endocarp cell layer disintegrates, while the mesocarp begins to dry out. It has been suggested that lignification may contribute to the silique shattering process, acting in a springlike manner to create mechanical tensions.

STAGE 19: The valves begin to separate from the dry silique, apparently owing to the lack of cell cohesion at the separation layer.

STAGE 20: At this stage the valves become separated from the dry silique and the mature seeds are ready to be dispersed.

Morphology, Histology and Development of Floral Organs

Sepals: In sepals L1 -derived cells form the epidermis, the mesophyll originates from the L2, and the L3 contributes to the vascu-lature in the basal part. Sepals and petals together form the perianth. Both organ types have a simple laminar structure, consisting of an epidermis, mesophyll and rather delicate vascular bundles (veins). The four sepal primordia (the abaxial, adaxial, and two lateral sepal primordia) are the first floral organ primordia to appear. They arise at stage 3 of flower development in a cruciform pattern. Whether all four sepals occupy one whorl or the two lateral sepals occupy a separate outer whorl, has been the subject of discussion, but all sepal primordia are formed at around the same time, shortly after they are specified.

The adaxial and abaxial surfaces of the sepal epidermis are different. On the abaxial surface, cells have irregular shapes and sizes with some quite long cells (with nuclei of various sizes) and fringes of smaller cells. Unlike the adaxial surface, the abaxial surface has stomata and may have unbranched trichomes.

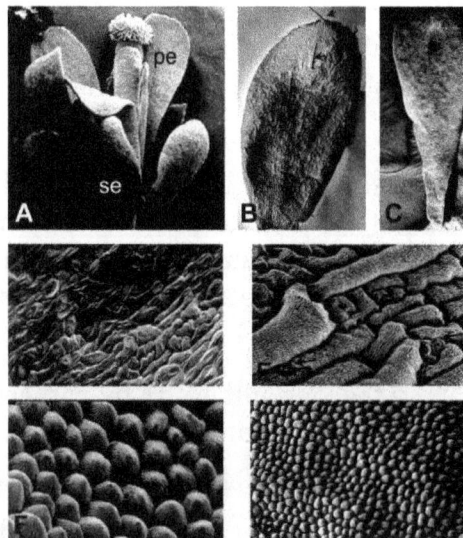

Sepal and petal cell types: Scanning electron micrographs (SEM)
of wild-type flowers and flower organs.

(A) A mature flower with sepals (se) and petals (pe) fully expanded and the stigma extending above the long stamens.

(B) Sepal blade showing simple unbranched trichomes (arrowheads) characteristic of the abaxial surface.

(C) Mature petal blade consisting of a basal claw and a distal blade.

(D) Adaxial sepal surface with irregular sizes and shapes of cells, some elongated (800x).

(C)Abaxial sepal surface bearing stomata (arrows) and characteristic elongated cells (500x).

(F) Adaxial surface of a mature petal blade showing conical cells with epicuticular thickenings running from the base to the apex (800x).

(G) Abaxial petal surface showing flatter, cobblestone-shaped cells with cuticular thickenings. Both petal surfaces lack stomata.

Petals: In the petal primordium the meristematic layer L1 contributes to the epidermis and L2 to the mesophyll; as yet cells originating from L3 have not been found to form part of the petal. These primordia become apparent almost at the same time as stamen primordia at stage 5 of flower development. Visible signs of petal differentiation are seen by stage 9. The four petals of Arabidopsis are white and flat and approximately the same size and shape. They are narrower and greenish toward the base.

Cells on the adaxial surface are conical with epicuticular thickenings running from the cell base to the apex, whereas those on the abaxial surface are flatter and more cobblestone-like with cuticular thickening. Stomata are absent from both petal surfaces. Cells toward the base of petals resemble those of stamen filaments.

Stamens: Primordia appear at stage 5 of flower development due to periclinal divisions in the subprotodermal cell layer (L2) and sometimes in L3. Stamen primordia are visible at stage 6. By stage 7, differentiation can be observed and long stamen primordia appear stalked at their bases. At this stage stamen primordia are composed of an L1 -derived epidermis, one layer of L2-derived subepidermis, and an L3-derived core. Locules appear in the anthers by stage 8. Growth of the internal anther tissue at this stage is due to divisions of L2-derived cells. At stage 14, anthers extend above the stigma. In the mature anther, the L3 cells contribute only to the vasculature. Stamens of the Arabidopsis flower are not formed simultaneously: four long medial stamens arise a little earlier than the two short lateral ones.

Each stamen consists of two distinct parts, the filament and the anther. At the tip of the filaments, the anther develops both reproductive and non-reproductive tissues that produce, harbor, and release pollen grains upon maturity. The anther is a bilocular structure with longitudinal dehiscence. Each locule develops from successive divisions of subprotodermal archesporial cells formed in the anther primordium that gives rise to three morphologically distinct layers: the endothecium, the middle layer, and the tapetum which surrounds the pollen mother cells (PMCs). The PMCs undergo meiosis and form the haploid microspores. The tapetum is a source of nutrients and is indispensable for microspore maturation. Anther development and microspore formation in Arabidopsis is a complex process that has been divided into 14 stages.

Once formed, PMCs are surrounded by a layer of callose. After meiosis, the anther contains most of its specialized cells and tissues, and tetrads of microspores are present within the pollen sacs; with microspores in each tetrad surrounded by a callose wall. Callose dissolves and microspores are released. As pollen grains develop, the anther enlarges and is pushed upward in the flower by the elongating filament.

Carpels: The fourth and innermost whorl is occupied by the gynoecium that is composed of two fused carpels. Carpel primordia start to form at stage 6 of flower development due to periclinal cell divisions in the L3 layer. Carpels enclose and protect the developing ovules, mediate pollination, and after fertilization develop into a fruit within which fertilized ovules develop into seeds. The gynoecium consists of two valves separated by a false septum with ovules arising from parental placental tissue on each side of the septum. The valves grow upward from the flower meristem to form a closed cylinder. At early stage 8, the walls of the cylinder are composed of an L1 -derived epidermis, one L2-derived subepidermal layer and a two-cell thick, L3-derived core. At this stage the distal L2 cells start to divide periclinally (with respect to the top surface of the cylinder), contributing to the longitudinal growth of the carpel. Later the inner surfaces of septal outgrowths within this cylinder will fuse, the tip will close and ovules will develop along the margins of the fused walls (placenta) of the bilocular chamber. The gynoecium is oriented in the flower so that the septum coincides with the medial plane.

At the distal end of the gynoecium, the stigma, an epidermal structure composed of stigmatic papillae (bulbous elongated cells), functions in pollen binding and recognition and participates in the induction of pollen germination. After germination, the pollen tubes will grow between the papillar cells into the transmitting tract at the center of the style and the septum of the ovary.

At about stage 11, the inner and outer integuments of the ovule are formed. By stage 12, the integuments of the developing ovule grow to cover the nucellus and megagametogenesis occurs.

Nectaries: These organs produce and secrete nectar. Nectar is a protein- and carbohydrate-rich solution, which varies in composition among different plant species. Nectar may be a reward for pollinators or for insects that protect the plant against herbivores, or even a lure for animal prey in carnivorous plants.

In Arabidopsis, the nectarium (multiple nectary) found in individual flowers is composed of two parts: nectary glands that form below the stamen filament, and the connective tissue linking the glands in a continuum around the androecium. The nectarium is always situated in the third whorl of the flower and its location is independent of the identity of the other organs occupying this whorl. These glands are formed from stage 9 to 17 of flower development.

Molecular Genetics of Arabidopsis Flower Development

Plant organogenesis, including flower formation, occurs from actively proliferating meristems over the entire life cycle.

Shoot Apical Meristem Proliferation and Maintenance

The balance between cell proliferation and cell recruitment to differentiated tissues in the SAM is dependent on mechanisms regulated by WUSCHEL. The homeodomain-containing WUS transcription factor has the role of the maintaining the identity of stem cells in the organizing center of the CZ; wus mutants lack stem cells in the SAM. WUS expression is limited to the cells immediately below the stem cells, an expression domain regulated by the receptor-kinase signaling system that includes the CLAVATA1, 2 and 3 (CLV1, 2, 3) gene products. CLV1 is expressed in most L3 stem cells while CLV3 is expressed in all three stem cell layers but mostly in L1 and L2 stem cells. In civ mutants, there is an imbalance between cells retained within meristems versus those recruited to form lateral organs, clv mutations cause an expansion of the WUS expression domain resulting in an enlarged stem cell niche. CLV3 expression is, in turn, positively regulated by WUS, suggesting that meristem size depends greatly on a WUS-CLV regulatory loop. Overexpression of CLV3 represses WUS expression and decreases meristem activity, suggesting that CLV3, a secreted CLE-domain peptide, is the signal that regulates WUS expression via the CLV1/CLV2 LRR protein-kinase transduction complex. It has been shown that other LRR-protein kinases closely related to CLV1 like BARELY ANY MERISTEM1 and 2 (BAM1, 2) are also involved in meristem maintenance possibly by sequestering CLV3 on the flanks of the meristem where they are expressed.

SHOOT MERISTEMLESS (STM) is a KNOTTED1-like homeobox (KNOX) gene that encodes a protein expressed in the SAM's CZ, RZ and regions of the PZ that have not been assigned to a primordium, i. e. it is expressed throughout the meristem except for anlagen, the sites of primordium formation. STM promotes the profileration of stem cell derivatives until a critical cellular mass is attained sufficient to form either leaves or floral primordia. It also inhibits the expression of ASYMMETRIC LEAVES1 and 2 (AS1, 2) genes in the SAM, preventing these cells from undergoing premature differentiation. Thus, the STM gene is considered to play a pivotal role in meristem maintenance. ULTRAPETALA1 (ULT1) encodes a cysteine-rich protein with a B-box like domain that restricts the size of shoot and floral meristems. It functions antagonistically to the proliferative roles of WUS and STM during most of the Arabidopsis life cycle but it in an independent genetic pathway.

Floral Meristem Specification and Determination

The changes in cellular characteristics, growth and geometry observed in the transition of the SAM to an IM are correlated with dynamic changes in the spatial and temporal expression of certain genes. The Arabidopsis IM produces rudimentary bracts in whose axils flower meristems emerge. STM and AINTEGUMENTA (ANT) expression patterns correlate with the development of this rudimentary bract primordium.

The expression of LEAFY (LFY), a transcription factor found only in plants and ANT has been used in order to trace the cells that form the flower primordium. First, tens of cells are rapidly recruited to those already committed to become part of the flower meristem. This stage may correspond to the upward bulging at the shallow crease formed between the rudimentary bract and the IM described by Kwiatkowska. These cells which express LFY then continue to proliferate. Interpretating this, the first cells that express LFY would correspond to the rudimentary bract (but not its axil or shallow crease), and later the domain of LFY expression would expand to include the cells committed to the flower primordium proper. This interpretation can explain the discrepancy in the number of founder cells estimated using sector boundary analysis and using in vivo LFY expression patterns. Bossinger and Smyth concluded that a FM derives from four founder cells directly on the surface of the IM (or SAM). In support of this, evidence from confocal laser scanning microscopy indicates that flower primordia are formed from two rows of cells in a radial arc. In contrast, the number of cells expressing LFY at these early stages suggest that a flower meristem has more founder cells. An explanation that resolves the discrepancy is that the LFV-expressing cells could include those that eventually form the rudimentary bract, as well as those which form the flower primordium.

The gene CUP-SHAPED COTYLEDON2 (CUC2) is expressed in the slow-dividing cells that expand in a latitudinal direction to define the second boundary between the floral primordium proper and the IM. Several regulators of CUC including a miRNA have been described as important components of the GRN involved in this developmental process.

Flower versus inflorescence meristem identity is controlled by a complex GRN that integrates environmental and internal cues. On induction of flowering, the IM genes, such as TERMINAL FLOWER and EMBRYONIC FLOWER 1 and 2, are repressed in the FM, while the floral meristem identity (FMI) genes, mainly LFY, APETALA1 (AP1), APETALA2 (AP2), and CAULIFLOWER (CAL), are upregulated.

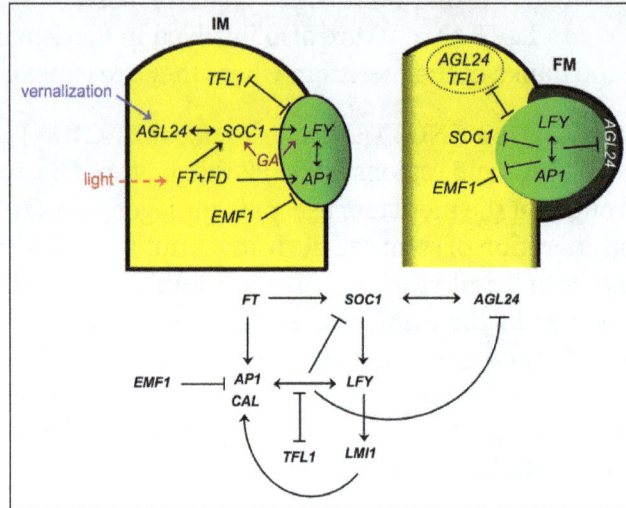

Inflorescence shoot apical meristem (IM) versus flower meristem (FM).

Simplified model of a gene regulatory network (GRN) that induces and maintains the FM. Flowering induction genes like FT, SOC1 and AGL24 are highly expressed in the IM in response to external (vernalization and light) and internal (gibberellins; GA) signals. These proteins in turn promote the expression of flower meristem identity (FMI) genes, LFY and AP1. Paradoxically, during the establishment of the FM, genes like TFL1 and EMF1 that help to maintain the IM identity are also expressed, keeping the expression of the FMI genes out of the IM. Later in development, LFY and AP1 repress the expression of TFL1 and flowering genes SOC1 and AGL24, among others, thus maintaining the FMI. Arrows and bars indicate positive and negative regulatory interactions respectively.

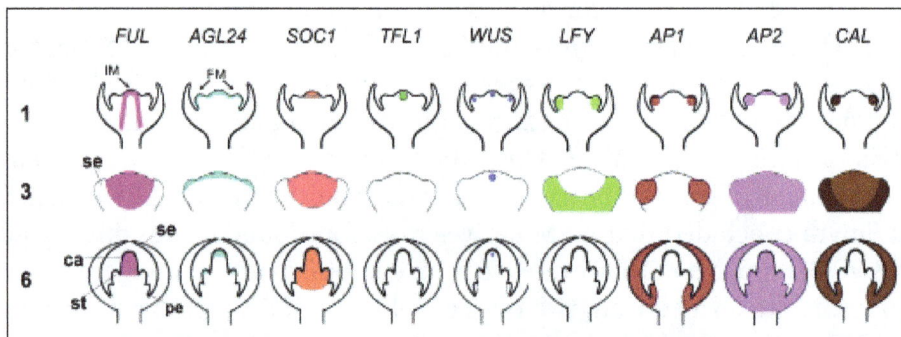

Schematic representation of some inflorescence shoot apical (IM) and flower (FM)
meristem gene expression patterns at stages 1, 3 and 6.

Flowering (FUL, AGL24 and SOC), indeterminate (WUS and TFL1), and FMI (LFY, AP1, AP2 and CAL) gene expression patterns based on in situ hybridization data during floral primordium developmental stages 1, 3 and 6. At stage 1, expression patterns correspond to their functions in IM and FM identities. Sepal (se), petal (pe), stamen (st) and carpel (ca) primordia are indicated.

At stages 3 to 6, all with the exception of TFL1 are expressed in the FM, probably because their respective proteins also affect organ development. FUL will participate in fruit development, LFY will induce all the ABC genes and AP1 and AP2 are fundamental in sepal and petal formation.

Mutual repression of the IM and FMI genes seem to underlie the co-existence, identity and boundaries of both types of meristem in the SAM in the transition to flowering. For example, if genes such as TFL1 or EMF1 or 2 are mutated, LFY and/or AP1 are ectopically expressed in the IM that is then transformed into a FM. On the contrary, if AP1, CAL and LFY are repressed, the FM attains IM identity. TFL1 is an important regulator of inflorescence development. It encodes a phosphatidyl ethanolamine-binding protein (PEBP) that is transcribed in the center of the IM but the protein moves to other cells where AP1 and LFY are downregulated. EMF genes are required for vegetative growth, but they seem to regulate flowering time and inflorescence development too. Loss-of-function mutants in these genes produce flowers immediately after germination skipping the vegetative phase. EMF1 encodes a transcription factor that represses AP1 but not LFY, and EMF2 encodes a novel zinc finger protein related to the polycomb group.

LFY is necessary and sufficient to specify FMI. In lfy mutants, leaves and secondary shoots are produced instead of flowers and LFY overexpression causes the conversion of leaves and axillary meristems to flowers. LFY is expressed in the leaf primordia during vegetative growth, but when induced by external (vernalization and light) and/or internal (gibberellins) signals, it is strongly expressed and relocates to the SAM flanks where floral meristems are formed. LFY expression persists at high levels in the FM until stage 3 of development and then diminishes in the center of the flower. LFY protein abundance, however, is homogenous in the FM, probably because it moves between cells.

LFY and AP1 have overlapping functions in establishing the FM; while the ap1 mutant has shoots with inflorescence characteristics, the lfy ap1 double mutant has an almost complete conversion of flowers into shoots. Both genes when overexpressed cause a terminal flower phenotype suggesting that each one is sufficient to determine the IM. CAL, the closest paralogue of AP1, and FRUITFULL (FUL) from the same gene clade within the MADS-box phylogenetic tree, may also act redundantly to AP1 in FM specification. Single cal and ful mutants do not show any FMI disorders, but in combination with ap1 in double or triple mutants, the ap1 phenotype is greatly intensified. FUL is expressed at the same time as LFY during the establishment of the FMI, but is mostly localized in the IM. Later during flower development, FUL is expressed again during carpel and silique development where it plays an important role. Despite its close similarity to AP1, overexpression of CAL is not able to determine the IM as does overexpression of AP1, indicating that CAL does not interact with the same partners as AP1. The unique functions of AP1 rely on residues within the K and COOH domains that are not found in CAL.

LFY directly regulates AP1 and CAL transcription by binding to the consensus sequence CCANTG. However, expression reminiscent of AP1 is seen in the lfy mutant, while it is completely abolished in the double mutant lfy ft (flowering locus t,). Thus FT, a homolog of TFL1 together with FD, a bZip transcription factor, redundantly regulate *AP1* with LFY. AP1 and CAL in turn regulate *LFY* by positive feedback, allowing it to exert its transcriptional regulation during flower development. Recently, additional LFY targets have been found, among them LATE MERISTEM IDENTITY1 (LMI1), which encodes a homeodomain leucine-zipper transcription factor and functions as a FMI gene. Interestingly, LMH acts together with LFY to activate CAL expression.

AP2 encodes a putative transcription factor of a plant-specific gene family (AP2/EREBP) with diverse functions. Mutations in AP2 enhance both ap1 and lfy mutant phenotypes, indicating that AP2 also plays a role in specifying FMI.

MADS-box genes are key components of the regulatory module that integrates flowering transition signaling pathways, IM and FM identities, and floral organ specification. To specify the FM, LFY and/or AP1 are also required to downregulate flowering induction genes such as AGAMOUS-LIKE24 (AGL24), SUPPRESSOR OF OVEREXPRESSION OF CO 1 (SOC1), SHORT VEGETATIVE PHASE (SVP), and FUL. Overexpression of any of these genes causes FM to revert to IM-like structures as when LFY and/or AP1 are mutated.

Floral reversion is often found in plants heterozygous for lfy6 (LFY/lfy) and homozygous for agamous-1 (ag-1), suggesting a key role for LFY and AG in the maintenance of determinate floral meristems. The reason for this is that late in floral organogenesis AG, induced by WUS, LFY and PERIANTHIA (PAN) among others, positively regulates KNUCKLES (KNU) which in turn represses WUS expression to terminate the stem cell niche after a limited number of organs have been formed. In fact, while WUS expression declines after stage 6 in wild-type flowers, it persists in pan or ag flowers. ULT also participates in meristem determinancy together with AG downregulating WUS.

Although it is very rare to observe spontaneous or induced reversion from FM to IM, a set of genes that actively maintain FM identity could conform to a "flower developmental module" that prevents reversion. The genetic mechanisms involved in maintaining FMI are closely linked to hormone balance and environmental factors. For example, we now know that STM is a positive regulator of local cytokinin (CK) biosynthesis and accumulation, and a repressor of gibberellin (GA) production. On the other hand, WUS enhances CK activity by repressing ARABIDOPSIS TYPEA RESPONSE REGULATORS (ARRs). The resulting high CK: auxin ratio and low GA levels promote indeterminate growth. While a high auxin concentration restricts STW and CUC expression, it also downregulates CK biosynthesis and activity, thus yielding a high auxin:CK ratio and high levels of GA, which induce floral meristem formation. Raising GA levels or response, for example by crossing with the spindly (spy) mutant, is sufficient to suppress FM reversion to IM in lfy, ap1, ap2 and ag mutants. This demonstrates the importance of GA in the maintenance of FM identity.

Light signal transduction pathways are also involved in FM maintenance. Spontaneous floral reversion in wild-type Arabidopsis has only been observed at low frequencies in the first flowers of Landsberg erecta grown in short days. However, long hypocotyl (hy1-1), a mutant in which photochrome activity is blocked, suppresses floral reversion of both lfy and ag single mutants in short days. Floral reversion seems to be a developmental abnormality with no apparent adaptive significance, unless plant resources are somehow saved under certain conditions if flowering is reversed. Further ecological and evolutionary developmental studies of Arabidopsis ecotypes will continue to elucidate the genetic, epigenetic, physiological, and environmental mechanisms involved in the maintenance of the FMI.

Specification of Floral Organs: The ABC Genes

Very soon after FM specification (11–13 days after germination in Landsberg *erecta* ecotype), the flower meristem is subdivided into four regions. Each one will give rise to the primordia of the different floral whorls, which from the outside to the inside are: sepals, petals, stamens, and carpels.

The genes responsible for floral whorl specification attain their spatio-temporal pattern as a result of regulatory interactions among themselves, interactions with meristem identity genes and with some other genes, such as WUS and UNUSUAL FLORAL ORGANS (UFO). The complexity of the interactions involved is shown in the 'floral organ specification gene regulatory network' (FOS-GRN) model. This model includes a set of interacting genes sufficient to pattern the IM and FM during the first stages of flower development.

One of the key FM identity genes is LFY. The protein encoded by this gene requires co-factors to set the spatial limits of expression of the floral organ identity genes AP3, Pl, and AG. For example, LFY participates with UFO in the regulation of AP1 and AP3 transcription, and with WUS co-regulates the expression of AG. LFY also regulates the expression of the SEPALLATA (SEP) genes SEP1, SEP2 and SEP3, additional MADS-box genes required for organ identity specification.

UFO is expressed in the second and third whorls during floral stage 3, probably restricting the B-gene expression domain to these whorls, together with LFY. The UFO gene encodes a protein containing an F-box domain, which is a characteristic of E3 ubiquitin ligases that are components of SCF (Skp Cullin F-box containing) complexes and mark proteins for proteosome-dependent degradation. It was recently shown that LFY interacts with UFO in order to directly bind the AP3 promoter. Furthermore, the proteosome activity mediated by UFO is required for the transcriptional activation of AP3 by LFY.

Key components of the GRN that underlies the early patterning of the flower meristem are the so-called ABC homeotic genes, AP1, AP2, AP3, Pl, and AG, which are all transcription factors belonging to the MADS-box gene family, except AP2.

The classic ABC model was inferred using Arabidopsis and Antirrhinum homeotic flower mutants. In these mutants two floral organ types are replaced by two other floral organ types as follows: A- class mutant flowers have carpels-stamens-stamens-carpels (from the outermost to the innermost whorl), B-class mutant flowers have sepals-sepals-carpels-carpels, and C-class mutant flowers have sepals-petalspetals-sepals. It was shown that mutations in all three functions lead to the transformation of all floral organs into leaf-like organs, suggesting that flowers are modified leaves. The Arabidopsis ABC mutants are shown in figure.

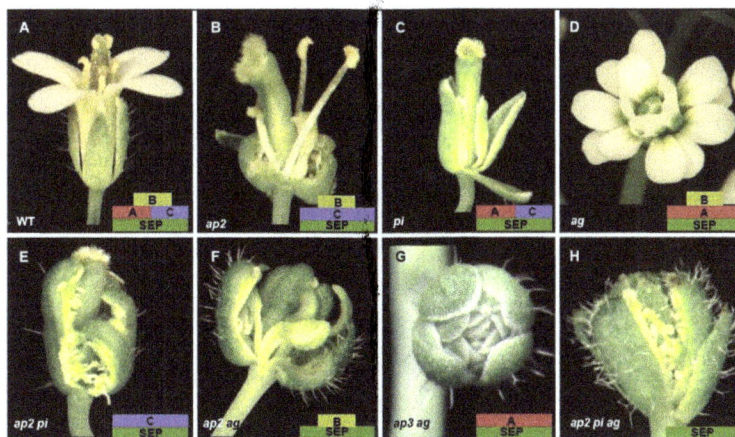

Arabidopsis ABC homeotic floral mutants.

Photos of single, double and triple ABC gene mutant flowers. Each photo is accompanied by a small diagram where rectangles represent the A (*AP1* and *AP2*), B (*AP3* and *Pl*), and C (*AG*)

combinatorial transcriptional regulatory functions and the *SEP* (*1, 2, 3, 4*) genes active in these mutants. Organs are listed below from the outer to the inner whorl unless stated otherwise.

(A) Wild-type (WT) flower.

(B) Single *ap2* mutant flower composed of carpelloid sepals, stamens, stamens and carpels.

(C) The *pi* mutant has flowers composed of sepals, sepals, carpels and carpels.

(D) The *ag* flower has the stamens transformed into petals and the carpels are replaced by another flower repeating the same pattern.

(E) The *ap2* pi double mutant displays flowers composed only of sepalloid carpels.

(F) The *ap2 ag* flowers have leaf-like organs in the first and fourth whorls and mosaic petal/stamen organs in the second and third whorls.

(G) The *ap3 ag* double mutants produce flowers composed of repeated whorls of sepals.

(H) The *ap2 pi ag* mutant has leaf-like organs with some residual carpel properties.

Hence, three different classes of homeotic genes with overlapping activities were proposed to be necessary for floral organ specification. The A function specifies sepals, the A and B functions specify petals, the B and C functions specify stamens and the C function specifies carpels. The A and C functions negatively regulate each other and the B function is restricted to the second and third whorls. The latter was originally thought to be independent of A and C functions, but it was later shown that the A function gene *AP1* regulates the B genes. AP1 binds to the promoter of *AP3*. AP1 can also specify petals by regulating the spatial domain of B genes together with UFO in the first flowers to arise, and independently of UFO in later flowers.

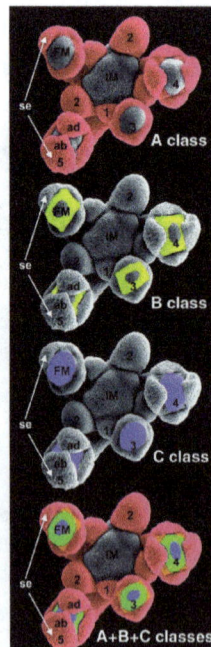

Expression patterns of the ABC genes during early stages of *Arabidopsis* flower development.

SEM of meristems have been colored to show expression patterns of A class (red, outer whorls), B class (yellow, petal and stamen primordia) and C class (blue, inner whorls) genes. Five flowers at early stages of development are marked 1 to 5 (5 being the oldest). Inflorescence shoot apical meristem (IM), floral meristem (FM) and sepals (se): adaxial (ad) and abaxial (ab) are indicated. .

Once identified at the molecular level, the mRNA expression patterns of the ABC genes were shown to overlap with the floral regions where the corresponding mutants had a phenotype. *AP1* and *AP2* are A-function genes. *AP1* is expressed in the two outer whorls of the floral meristem and is important for the establishment of sepal and petal identity as well as the FM. *AP1* expression is first up-regulated by LFY and FT/ FD, but later is maintained by the B class genes in a positive feedback loop. Strong *ap1* alleles (*ap1-1*) often lack petals in the second whorl, while weaker mutant alleles of this gene do not have a full homeotic conversion of floral organs.

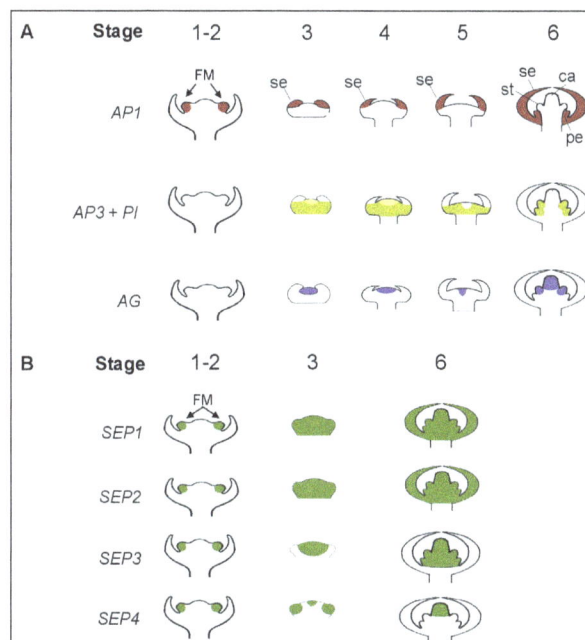

Diagram illustrating mRNA expression patterns of Arabidopsis
ABC and SEP genes during different stages of flower development.

(A) ABC gene expression patterns illustrated from stage 1 to 6. The A function gene *AP1* is expressed (red) in the two outer floral primordia whorls that will later develop into sepals (se) and petals (pe). The A function gene *AP2* is expressed in all four whorls of the flower. The B function genes (dark yellow) *AP3* and PI are expressed from stage 3 in the next two inner whorls of the flower. Interestingly *Pl* is also expressed at stages 3 and 4 in cells that will generate the fourth whorl (light yellow). After stage 5, the pattern of PI expression largely coincides with that of *AP3*, only in petal and stamen (st) primordia. The C function gene AG is expressed (blue) in the two inner whorls that will become the stamens and carpels (ca).

(B) SEP gene expression pattern during several stages (1 or 2, 3 and 6) of flower development. *SEP1* and *SEP2* are expressed in all whorls of the flower. *SEP3* is first detected in late stage 2 flower primordia and afterwards in petal (pe), stamen (st), and carpel (ca) primordia. The expression pattern at stage 6 was deduced that from at stage 7. *SEP4* is weakly expressed in sepal primordia at stage 3 and strongly expressed in carpel primordia from stage 3 to 6. Both figures have been modified and expanded from Krizek and Fletcher.

In contrast to the MADS-box ABCs, the expression pattern of AP2 does not correlate with the site where it exerts its function in floral organ identity. AP2 mRNA is found throughout the flower meristem. The AP2 is repressed at the translational level by a microRNA (miR172), which is active only in whorls 3 and 4, thus explaining why the function of AP2 is restricted to the first two whorls of flower organs. In a recent experiment using double mutants of ag and an ap2 allele, which is insensitive to repression by miR172, it was shown that both AG and miR172 independently downregulate AP2, but miR172 is more important than AG. ap2 mutants rarely develop petals and their sepals are transformed into carpelloid structures due to ectopic AG expression, which is negatively regulated by AP2 itself. AP2 is also implicated in the upregulation of the B genes, AP3 and Pl.

The B class genes (*AP3* and *Pl*) are expressed in the second and third whorls and mutant flowers of any or both of these two genes lack petals and stamens, as predicted in the ABC model. The fact that both single mutants yield the same phenotype shows their interdependence. *AP3* and *Pl* are regulated in two steps: they are first induced by LFY/UFO in response to flowering signals and they later maintain their expression in a self-regulatory loop. The proteins encoded by these two genes form heterodimers to exert their B function during petal and stamen development and this oligomerization is necessary for them to move into the nucleus.

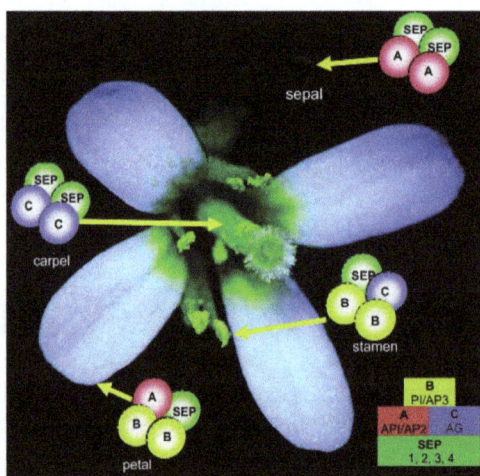

Schematic representation of the interaction of ABC and SEP
proteins in the quartet model for *Arabidopsis* floral organ specification.

Possible MADS-domain protein complexes (circles) of the ABC model are sufficient for the specification of each of the four floral organs. In the ABC model, rectangles represent the A (AP1 and AP2), B (AP3 and Pl), and C (AG) combinatorial transcriptional regulatory functions necessary for sepal, petal, stamen and carpel primordia specification. The green rectangle below represents the SEP (1, 2, 3 and 4) proteins that interact with proteins encoded by the ABC genes (unknown for AP2 which has not been tested) to specify each floral organ.

Both genes are also regulated positively in a regulatory loop by AP1 and negatively by EARLY BOLTING IN SHORT DAYS (EBS), a gene that encodes a nuclear protein that participates in petal and stamen development and regulates flowering time by repressing FT ANT, a member of the AP2 gene family, is another regulator of the B function, positively inducing AP3.

The only C-type gene discovered up to now is the MADS-box gene *AG*. *ag* mutant flowers lack stamens and carpels, and also bear indeterminate flowers with reiterating sepals and petals, suggesting that *AG* is important for floral meristem determinacy, besides its role in stamen and carpel identity. The regulation of *AG* has been much studied; at least ten proteins repress and five activate it to maintain its expression in the appropriate whorl.

AG is repressed by a transcriptional co-repressor complex formed by LEUNIG (LUG) and SEUSS (SEU). LUG encodes a transcription protein similar to TUP1 from yeast and interacts with SEU, which encodes a plant specific protein. Neither of these proteins are able to bind DNA sequences and AP1 and SEP3 recruit SEU/LUG to the second intron of AG to perform their inhibitory function and prevent the ectopic expression of AG. Recently, another transcriptional repressor of AG was identified, LEUNIG_HOMOLOG (LUH). This gene is the closest homolog of LUG and its inhibitory function on AG is completely dependent on SEU.

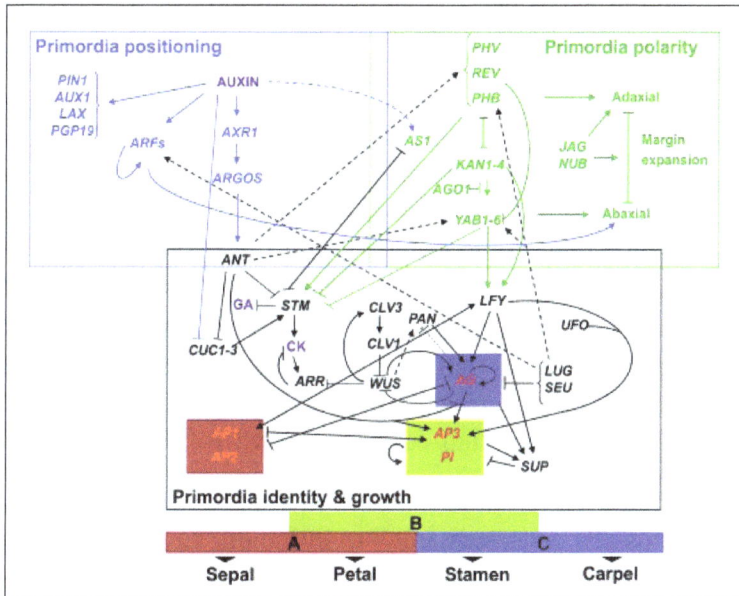

Functional gene regulatory modules during early flower development.

Common molecular modules act during early meristem morphogenesis from the SAM both before and after reproduction. During floral organogenesis, these modules interact among themselves and with the FOS-GRN that includes the floral homeotic genes. Anlagen positioning in the SAM flanks depends on auxin gradients. Transport and signal transduction proteins, as well as other factors regulated by auxins (letters in blue), participate in the establishment of such gradients and finally determining the position of primordia. The auxin pathway also downregulates some members of the NAC family (*CUC1* to *3* are important for organ boundary establishment), which also participate in the positive regulation of *STM* and *KNOX* genes. Since *WUS* maintains the apical meristem stem cells in a proliferating state with CLV proteins that in turn regulate its expression in a non cell-autonomous negative-positive feedback loop, and *STM* prevents meristem cell differentiation by inducing the production of cytokinins (CK) and the *ARR* transduction pathway, floral primordia may emerge if cells in the anlagen are able to downregulate *STM*. This can be achieved by the action of AS1 and ANT. Upregulation of *LFY* by the flowering genes in conjunction with some KAN and YAB proteins, activate the expression of ABC homeotic genes (in red) for the establishment of the floral organ primordia identity and growth. Lateral organ primordia acquire apical/basal, lateral/medial and adaxial/abaxial polarities by the action of protein families that include PHABs (PHB, PHV and REV), KANs (KANADI1–3, ATS/KAN4), YABs (FIL/YAB1, YAB2, YAB3, INO/YAB4, YAB5 and CRC/YAB6), JAG and NUB (letters in green). Some of these are organ-specific while others are shared by different floral organ primordia. Not all the genes involved in each module are depicted, just some of the most representative ones, which help us to understand how they are interconnected. Arrows and bars indicate positive and negative regulatory interactions, respectively, and dashed lines a postulated interaction not yet proven. The text color used for the gene names in each module is the same as in figures and where specific organ developmental processes are summarized and the ABC genes are shown. Hormones are in purple.

Another repressor of AG is BELLRINGER (BLR), a homeodomain protein that binds to regions in the second intron of AG and prevents ectopic AG expression in the two outer whorls of the flower. AG is also negatively regulated epigenetically by a histone acetyltransferase GCN5. Other genes that participate in floral organogenesis are repressors of AG, namely RABBIT EARS (RBE), ANT and STERILE APETALA (SAP). AG is also positively regulated at the post-transcriptional level by several ENHANCER OF AG-4 (HUA) and HUA ENHANCER (HEN) genes. All of these genes play a major role in pre-mRNA processing of AG.

The ABC proteins exert their regulatory function as multimers. In *Antirrhinum majus,* a ternary complex between A and B function proteins was found to bind CArG DNA boxes more efficiently than single proteins. More specifically, a higher-order complex consisting of SQUAMOSA (*SQUA,* the *AP1* ortholog), DEFICIENS, and GLOBOSA (*DEF* and, *GLO* are *A. majus AP3* and *PI* orthologues, respectively) bound DNA more efficiently than DEF/GLO or SQUA alone. These results suggest that transcriptional complexes that combine A and B function proteins are more stable than those formed with proteins of any one function alone.

The fact that the ABC genes are necessary but not sufficient to determine floral organ identity was later confirmed in Arabidopsis. Honma and Goto used a yeast three-hybrid method to show that SEP3 and AP1 are able to interact with the heterodimer AP3/PI but not with AP3 or PI alone. Moreover, they described this interaction as essential, since the heterodimer AP3/PI lacks the activation domain necessary for a transcription factor to function, a domain which both SEP3 and AP1 possess. These findings suggest that the inclusion of SEP3 or AP1 together with AP3/PI could result in an active tetrameric transcriptional complex. It was also demonstrated that the ABC proteins on their own or combined according to the ABC model (A, AB, BC, or C) were not sufficient to determine floral organs when expressed in leaves under the action of the 35S constitutive promoter. However, floral organs could indeed be recovered in leaves once appropriate combinations of genes were expressed.

The SEP genes received their names because the floral organs that develop in all four whorls in triple sep mutants resemble sepals. This sep1 sep2 sep3 triple mutant phenotype is markedly similar to that of double mutants that lack both B and C class activity, such as pi ag and ap3 ag in which the floral meristem becomes indeterminate as well. Single or double mutants for these SEP genes yield flowers indistinguishable from wild type, thus suggesting that the three SEP genes are functionally redundant and are important in determining three of the four floral organs: petals, stamens, and carpels.

The triple sep1 sep2 sep3 mutant does not show alterations in sepal identity, an additional gene is likely to be involved in sepal specification. Indeed, another SEP-like MADSbox gene, SEP4, has now been characterized, and the quadruple sep1 sep2 sep3 sep4 mutants produce flowers with leaf-like organs in all whorls, thus confirming the SEP genes contribute to each floral organ identity. Coincidently, SEP genes are expressed in the whole floral meristem during flower development, are important in regulating B and C gene expression, and encode proteins that apparently interact with the ABC proteins.

Based on data from Antirrhinum and yeast two-hybrid and three-hybrid protein interactions, and on the phenotypes of the ABC mutants, three models have been proposed to explain how the MADS domain proteins interact to constitute functional transcriptional complexes and bind DNA. None of the models completely explains the experimental data available, but the quartet model seems the most plausible. This model proposes that MADS domain proteins form tetrameric complexes during floral organ determination. Within each transcriptional complex, there would be two MADS dimers, each one binding to a single CArG-binding site causing the DNA of the promoter region to bend, enabling the MADS dimers to act cooperatively in a tetrameric complex to regulate the gene. For example, binding of one dirtier within the tetramer to DNA could increase the affinity of the second dirtier for local DNA binding. Besides, one of the dimers could function as the activation domain of the tetramer allowing for efficient transcriptional activation. Interestingly, several dimers and potential tetramers have been documented in a complete Arabidopsis MADS-domain family protein-protein interactome via yeast two-hybrid interactions. This data base has been updated with a yeast three-hybrid screen for MADS-domain proteins.

Target Genes of the ABCs

Target genes of the ABC genes link the floral organ specification - gene regulatory network (FOS-GRN) with processes in organ primordia establishment and development. Among the direct targets of the ABC genes, transcriptional regulators and hormone-related genes are prominent. But the sets of target genes change as organ development progresses; at later stages of floral organ development, several components of what could be common modules have been found that are involved in generic developmental processes during sepal, petal, stamen and carpel development. Finally, multiple genes having cell-specific roles are turned on especially during stamen and carpel development that is much more complex than perianth development.

The first examples of genes regulated by the ABC genes were two MADS-box genes, AGL1 and AGL5 (renamed the SHATTERPROOF genes (SPH1 and 2, respectively). There is virtually no expression of either gene in ag mutants. SHP2 is only expressed in carpels and AG can activate an SPH2:GUS reporter construct; furthermore, AG binds its promoter in vitro.

The first non-ABC gene identified as a target of a MADS A, B or C protein was NAC-LIKE ACTIVATED BY AP3/PI (NAP), a target of the AP3/PI complex. It is important for the transition between cell division and cell expansion during petal and stamen development.

The two genes negatively regulated by AP3/PI, GATA NITRATE INDUCIBLE, CARBON-METABOLISM-IN-VOLVED (GCN) and GCN-LIKE(GNL), a GCN paralog. Both genes regulate chlorophyll biosynthesis in plant cells. Thus, their downregulation could be important in preventing chlorophyll accumulation in petals and anthers. The same shows that both GNC and GNL, together with the B class genes, regulate the expression of a number of other GATA-motif-containing target genes like HEXOKINASE1 (HXK1,).

SUPERMAN is upregulated by AP3/PI and AG and by LFY. SUP encodes a transcription factor with a C2H2-zinc finger motif and is involved in the maintenance of the stamen/carpel whorl boundary. While the B genes and LFY seem to regulate early SUP expression, AG and the B genes are involved in maintaining its expression in flowers from stage 5 onward.

Recent microarray experiments have proved useful in revealing new targets of the ABC MADS homeotic genes, as well as many putative components of the complex networks involved in floral organogenesis. For example, it was shown that the AP3/PI dimer regulates, directly or indirectly, 47 target genes. Only two of these are transcription factors, while most participate in basic cellular functions required for stamen and petal development. By contrast, AG controls, directly or indirectly, the expression of 149 genes most of which are transcription factors, including other members of the MADS-box gene family. Ten of these were also shown to be direct targets of AG using chromatin immunoprecipitation (ChIP), including AG itself, AP3, CRC and ATH1, a gene that encodes a BELL-type homeodomain protein that participates in the development of the basal region of shoot organs.

A more exhaustive experiment used four homeotic mutants (ap1/ap2, ap3, pi and ag) in two types of microarray assays: a whole genome microarray with approximately 26, 090 gene-specific oligonucleotides and a flower specific-cDNA microarray with 5, 000–6, 000 genes. To summarize the assay results, transcription factors were neither over or underrepresented as being regulated by the ABC genes; on the contrary, genes involved in general cellular maintenance (DNA recombination and protein synthesis) were underrepresented. Genes specifically expressed in each of the four different whorls were identified: 13 genes for sepals, 18 for petals, 1162 for stamens, and 260 for carpels. As expected from their structural and cellular complexity, the reproductive floral organs had many more specific target genes than the perianth organs.

In another genomic study of early floral stages it was found that many genes were down regulated in incipient floral primordia while many of them were activated during the differentiation of floral organs. However, some genes were overrepresented during all stages analyzed (i. e. transcription factors including the family of MADS-box genes, PIN dependent auxin transport genes, as well as auxin and GA metabolism genes). Even though the MADS box genes were overrepresented, the promoter regions of the genes expressed during these different stages are not enriched in CArG-box sequences compared to random samples from the whole genome. This result suggests that MADS-domain transcription factors may be able to bind sequences other than CArG motifs, or that they have few direct targets during the developmental stages analyzed.

In a different approach, an inducible post-translational version of AG was used in gene expression profiling to detect AG target genes. One of the genes identified that is upregulated by AG is SPOROCYTE-LESS (SPL). AG is able to bind in vitro to the 3' region (downstream of the stop codon) of the SPL gene. SPL has been described as a key regulator of sporogenesis later during stamen and carpel development.

Floral Organogenesis

The challenge of inferring the topology of the gene regulatory network (GRN) underlying the establishment of floral organ primordia, and their development (cell differentiation, morphogenesis and growth) is still ahead. Such modules involve several functional feedback loops and underlie different generic developmental processes mainly: primordia type specification; delimitation; floral organ primordia positioning that depends on fundamentally on auxins; primordia number; inter-whorl and within-whorl boundaries; and primordia and organ adaxial-abaxial polarity. At later stages of floral organ development, subcellular differentiation and patterning, as well as overall organogenesis takes place and more specific regulatory modules are involved. The genes within such modules are treated separately for each organ type.

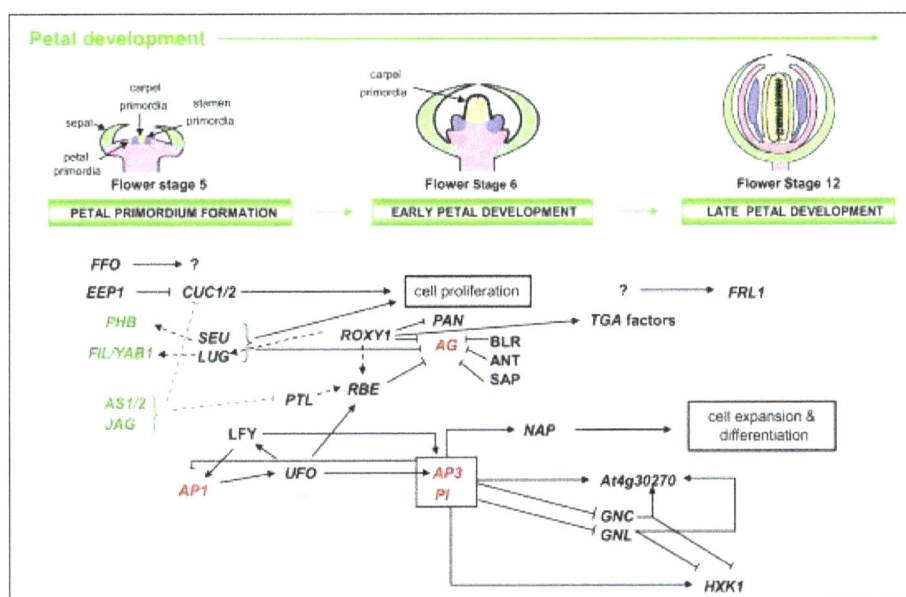

Main stages of petal development and some genes involved.

Schemes at the top illustrate three different stages of petal development. Briefly, GRN modules (genes) in petal development include those involved in the establishment of the second whorl domain, the specification of petal identity and cell differentiation. CUC genes under the regulation of miR164c are involved in establishing whorl boundaries. Genes involved in polarity determination like JAG, PHB and YAB1 are also necessary for petal development. A, B and SEP genes, and the absence of C genes, determine petal identity (AP2 and SEP genes are not shown here for clarity). Petal blades are formed by active cell division at early developmental stages and by cell enlargement and differentiation at later stages. Some of the genes expressed early need to be continuously expressed throughout petal growth, including ROXY1, SEU, and LUG. Downregulation of the GNC, GNL, and HXK1 genes inhibits chlorophyll accumulation and expression of photosynthetic genes. At4g30270 might be necessary for correct cell wall dynamics during petal growth. Gene color code as in figure; arrows and bars indicate positive and negative regulatory interactions, respectively.

Schemes of some stages of flower development showing representative stages of anther cell differentiation are shown at the top. At stage 1 of anther development and microspore formation, rounded stamen primordia emerge with three cell layers, L1, L2 and L3. During stage 2, the archesporial cells (Ar) arise in the four "corners" of the L2 layer and the epidermis in the L1. Before meiosis the Ar cells divide and generate the primary parietal layer (1°P) and the primary sporogenous layer (1°Sp). The 1°P then divides into two secondary parietal layers (outer and inner, 2°P). The outer layer gives rise to the endothecium, the inner cells to the middle layer and the tapetum. 1°Sp produces the microspore mother cell (MMC) that undergoes meiosis and gives rise to the

microspores. At stage 7, meiosis is completed and the four locules carrying tetrads (Tds) of microspores are seen. At stage 14, cells shrink and the anther dehisces liberating the pollen grains. Some
of the known genetic interactions important during anther development are shown in purple. AG
(in red) induces the expression of SPL (the first gene known to be committed to anther development); later during microsporangium formation the action of the EMS, DYT, MS1 and AMS genes
is also indispensable. Arrows and bars indicate positive and negative regulatory interactions, respectively, and dashed lines possible indirect interactions.

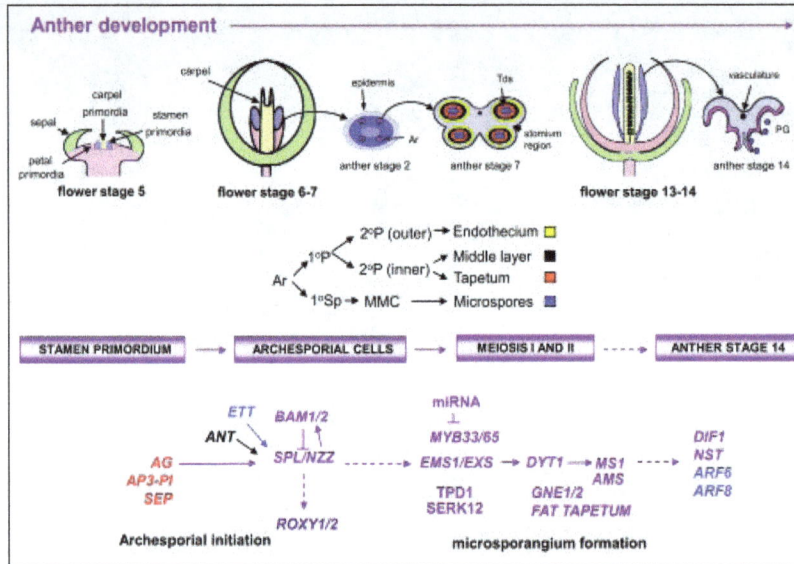

Stages of stamen development with emphasis
on the genes implicated in anther formation.

As a precursor to integrating GRN modules in the above categories, we now provide a synthesis of
the molecular genetic studies of how such generic developmental processes are regulated. Several of these have also been identified as important regulators of leaf development, substantiating
the proposal of Goethe that all plant organs are elaborations or modifications of a core leaf-like
developmental program. ABC floral organ identity genes are also important in fine-tuning or co-
ordinating the role of genes involved in some of the generic developmental modules during flower
development. Some genes participate in more than one process or module and are important for
making connections between different GRN modules. In such cases, they are considered in more
than one category.

Regulatory modules controlling distinct components of floral organ development have been eluci-
dated to different extents depending on available mutant phenotypes. In correlation with anatomical and morphological complexity, the size and complexity of the regulatory modules underlying stamen and carpel development are much greater than those that regulate sepal or petal development.

In the flower meristem, normal organogenesis depends upon a homeostatic equilibrium between
stem cell specification and cellular differentiation. Plant morphogenesis is influenced both by the
orientation and rate of cell division, as well as by cell expansion and differentiation. How the molecular aspects of these processes are coordinated has been very difficult to elucidate. However, it
is generally accepted that cells in meristematic regions respond to positional information important for inducing and controlling morphogenesis. One of these positional signals is auxin. Several

mutations that affect the number, size, and/or shape of one or several floral organs have also been characterized. Some of these phenotypes are pleiotropic consequences of mutations in genes acting from earlier steps of plant and flower development. Others are the result of alterations in organ specific genes.

Floral Meristem and Organ Primordia Positioning: The Role of Auxin

The shoot apical meristem produces leaves and then flowers in a highly predictable and regular phyllotactic pattern. One of the key compounds that regulate this developmental process is the hormone auxin. Increased auxin levels mark the initiation sites for organ primordia (including those of floral organs) and local application of auxin is sufficient to trigger leaf or flower formation in the shoot apex. Once the primordium is established, there is a depletion of auxin around it and another peak of auxin is only able to form in cells at a specific distance from pre-existing primordia, generating a phyllotactic pattern. After Initiation, the primordium grows by cell proliferation and cell expansion, and the organ differentiates along the apical-basal and dorsal-ventral axes.

The overall distribution of auxin depends on its biosynthesis, metabolism, and directional transport. Most auxin is synthesized in young tissues of the shoot and distributed throughout the plant by two physiologically distinct pathways. One of them is passive and occurs only by diffusion through the mature phloem. The other one is an active polar auxin transport (called PAT) that mediates cell-to-cell movement of auxin through two different types of proteins, efflux and influx carriers. Some of the genes that encode these transporters (or carriers) have been cloned: PIN-FORMED (PIN) and P-GLYCOPROTEINS (ABCB/PGP) for auxin efflux, and AUXIN1 (AUX1) and its paralogs LIKE-AUX1 (LAX1-3) for auxin uptake/influx.

The PIN gene family encodes eight protein members in total; three of them (PIN5, 6, and 8) of unknown function. All of the PIN proteins characterized until now are asymmetrically distributed on the plasma membrane and some of them can be found in specific cell types with no pronounced polarity. The direction of auxin flow is believed to be determined by the asymmetric cellular localization of PIN proteins. The first of these proteins to be characterized was PIN1, and its mutation (pin1) results in pin-shaped inflorescence meristems without flowers. PIN1 expression is induced by auxin and it encodes a protein with 10–12 putative transmembrane domains and shares some similarity with bacterial transporters. pin1mutant plants accumulate high amounts of auxin in vegetative meristems and a deficiency in the apical inflorescence meristem, which results in a defective organ initiation of leaves and flowers, a phenotype that can be imitated in wild type using auxin efflux inhibitors. Of the other PIN proteins, only pin3 and pin7 loss-of-function mutants have flowers, and these bear fused petals, no stamens, and occasionally no sepals. PIN3 is essentially involved in mediating differential shoot growth and PIN7 is important during early embryo development.

Auxin movement mediated by PIN carrier proteins determines the growth axis of the developing organ by establishing an auxin gradient with its maximum at the tip. As the primordium rapidly expands, auxin is depleted from the tip. Two hypotheses have been proposed to explain this observation: either auxin is transported through the primordium interior into the vascular network or it is depleted from primordial regions as a result of specific reversals in PIN1 polarity.

The ABCB/PGPs are also transmembrane proteins that belong to the ATP-binding cassette (ABC) transporter superfamily. In Arabidopsis, three of their members, ABCB1, ABCB4, and ABCB9, are able to transport auxin away from apical tissues and are important in maintaining long-distance auxin transport. One of the PGP proteins (PGP19) co-localizes and interacts with PIN1 and the ABCB protein is apparently important in stabilizing plasma membrane microdomains necessary for enhanced PIN1 activity.

Auxin enters the cell passively by simple diffusion and also by the import activity of AUX1 and related LAX proteins. The *AUX1* gene encodes a protein with 11 putative transmembrane domains similar to plant amino acid permeases. The mutant form (*aux1*) was identified in a screen for auxin resistant and agravitropic mutants. The AUX1 protein also has polar subcellular localization in some cells and co-localizes with PIN1 in the shoot apical meristem. AUX1/LAX function could be essential for stabilizing the phyllotactic pattern. The proposed model for AUX1/LAX function is that these proteins concentrate auxin in the cytoplasm of cells of the L1 layer, preventing auxin diffusion in the apoplast.

PINOID (PID) encodes a Ser/Thr protein kinase which has been implicated to function in redirecting subcellular PIN polarities, because the loss of its activity causes a shift in apical-basal PIN polarity. pid mutants have a defect in organ formation similar to that of pin1, but they do produce a few flowers with altered floral organ numbers (more petals but fewer stamens). Recently, Michniewicz et al., (2007) reported that In vivo PIN1 phosphorylation is directly dependent on the kinase PID and a phosphatase PP2A, which may act directly by dephosphorylating PIN1 or indirectly through PID. This phosphorylation status determines the intracellular apical-basal localization of PIN1 and therefore auxin transport-dependent development. PIN1 is targeted to the apex when it is phosphorylated and to the base when it is dephosphorylated.

Accumulation of auxin activates downstream processes through specific receptors and the combinatorial action of members of two large families of transcription factors, AUXIN RESPONSE FACTORS (ARF) and IAA/AUX. The Aux/IAA proteins are degraded when the levels of free auxin rise, resulting in derepression of ARFs. ETTIN (ETT)IARF3 has a dynamic role in patterning by acting in specific cells within floral meristems and reproductive organs. At early stages, ETT functions in determining the number of organ primordia, whereas later it is involved in the outgrowth and patterning of tissues within organ primordia. err mutant plants show altered flower development; some flowers have missing petals and rudimentary radialized stamens, and others have normal fertile stamens, but radialized petals. ETT is also involved in prepatterning apical and basal boundaries in the gynoecium primordium. MONOPTEROS (MP)/ARF5 mutants (mp) have inflorescences with smaller or absent flowers, similar to pin1 mutants.

Floral Organ Primordia Number, Size and Boundaries

In Arabidopsis, which is a self-fertilizing (autogamous) and partially cleistogamous (before flower bud opens) plant, floral organ size might not be under strong evolutionary pressure compared to allogamous species. However, it has been an important model to study genes that control size and architectural traits of flowers.

Several mutations that affect meristem size and maintenance lead to alterations in flower organ number or size. Mutations in the CLV genes cause an increase in meristem size, thus yielding additional whorls and a change in floral organ number with altered phyllotaxis. Mutations in genes that control cell proliferation in the SAM, such as the CLV genes, are similar to ULT in that they have larger SAM and primordial.

When WUS is repressed and the number of cells for floral primordia formation is reduced, organ architecture is compromised suggesting that there is a threshold number of cells required to form a normal organ. In fact, the loss of organs observed in Á-function mutants, or any other AG repressor mutant could be explained as a result of premature repression of WUS by AG in these organs.

Other mutants that have altered floral organ numbers are pan, ett and sup. Both pan and ett have more sepals and petals and fewer stamens, whereas sup produces more stamens at the expense of carpels. Double pan sup mutants however have an attenuated sup phenotype in the fourth whorl, probably because in this mutant AG is downregulated and the domain of expression of WUS is expanded.

The PAN gene mutation specifically alters floral organ number, yielding fertile plants with a pentamerous meristic pattern. PAN encodes a member of the bZIP class of transcriptional regulators and is thought to act in the process by which cells assess their position within the developing floral meristem. This gene may affect the switch that commits floral organ primordia cells to enter an organ initiation program. PAN and WUS expression overlaps and in clv mutants both genes are ectopically expressed. WUS overexpression causes PAN overexpansion too suggesting that this gene is positively regulated by WUS.

Interestingly, pentameric symmetry is characteristic of flowers in early-diverging angiosperm lineages, thus suggesting that PAN may have been involved in changes to meristic patterns during angiosperm diversification; particularly the evolution from pentamerous to tetramerous flowers in the Brassicaceae lineage.

Organ size is also regulated by the same components in all whorls. The ANT gene encodes a transcription factor of the AP2 family, which seems to be a general regulator of organ size during organogenesis. The overexpression of ANT causes increased cell division in sepals and increased cell expansion in the inner three whorls, probably affecting both the rate and duration of cell divisions which are important determinants of the final size of lateral organs. ARGOS participates in the same transduction pathway as ANT and acts downstream of AUXIN RESISTANT 1 (AXR1). Interestingly, increased organ size observed in ARGOS overexpression lines is due to an extended period of cell division rather than to an increase in growth rate. So, it is plausible to assume that these two genes (and probably others) affect organ size by transducing signals from plant growth regulators, such as auxin, which is a key player in establishing SAM primordia and a general regulator of cell proliferation and expansion.

ANT also participates in defining abaxial-adaxial organ polarity in combination with FILAMENTOUS FLOWERIYABBY1 and thus may be one of the links between the modules controlling primordia growth and the polarity establishment.

Ectopic expression of UFO also causes increased floral organ size, due to increased cell division. This pathway is regulated by UFO independently of its role in B gene expression, because ectopic expression of the B genes does not induce any increase in organ size, so missexpression of other unknown UFO-dependent factors may account for this phenotype. UFO and two gene enhancers of the ufo⁻ phenotype, FUSED FLORAL ORGANS 1 and 3 (FFO1 and FFO3), could also participate in establishing and maintaining organ boundaries probably by affecting cell proliferation.

Morphological boundaries are established in the early stages of the formation of a primordium separating it from surrounding tissues, and later from adjacent organ primordia. Cells in the boundary are distinctly narrow and elongated with low proliferation rates. Genes expressed in the boundary may affect both meristem and organ development by upregulating cell differentiation genes and downregulating meristematic genes. CUC1, 2, and 3 encode NAC-domain transcription factors that promote morphological separation of lateral organs through growth repression. cuc1 cuc2 double mutant seedlings have fused cotyledons with no shoots. However, when adventitious stems are induced in this genotype, flowers have fused sepals and stamens, fewer petals and stamens number, and reduced fertility. CUC genes are epigenetically regulated.

Other genes, such as LATERAL ORGAN BOUNDARY (LOB) and JAGGED LATERAL ORGANS (JLO), members of the LATERAL ORGAN BOUNDARY DOMAIN (LBD) gene family, encode putative transcription factors with a leucine-zipper motif that are also expressed in boundary cells. JLO along with the CUC genes probably coordinate auxin accumulation and loss of meristem-specific gene expression in organ anlagen.

Floral Organ Polarity

Establishing organ polarity is an important aspect of morphogenesis and it is sometimes clearly associated with specific functions of plant organs. Both, adaxial-abaxial and proximal-distal polarities are regulated by genetic circuits that are similar for all lateral organs, although each organ type has distinct cell types and morphogenesis in the abaxial versus adaxial surfaces, and in the proximal versus distal regions. Organ polarity is also linked to the establishment of hormone gradients.

Briefly, abaxial fate is conferred by members of the YABBY family and by some of the KANADI genes, whereas adaxial cell fate is determined by members of the PHAB family: REVOLUTA (REV), PHABULOSA (PHB), and PHAVOLUTA (PHV).

YABBY proteins (YAB) are transcription factors with a Zn-finger and a helix-loop-helix (YABBY) domain that are promoters of abaxial cell fate in all lateral organs, among other functions. During flower development they participate in establishing the primordium domain and meristem patterns, and later in maintaining abaxial polarity. FIL/YAB1, YAB2, and YAB3 are expressed in a polar manner in all lateral organs of the flower meristem, while CRABS CLAW(CRC/YAB63) is only expressed in carpels and nectaries, and INNER NO OUTER (INO/YAB4) is restricted to outer integuments.

KANADI (KAN) genes encode transcription factors of the GARP family. KAN1, KAN2, and KAN3 have been implicated in promoting abaxial cell fates. The kan1 mutant was selected as a genetic enhancer of crc gynoecium phenotype, producing a mirror-image of adaxial tissues in the kan1 crc double mutant, indicating that both genes participate in a redundant manner to promote abaxial identity. In kan1 kan2 double mutants, all floral organs are also extremely adaxialized. Although these KAN genes are not necessary for the activation of YAB genes, they are important in controlling their proper abaxial localization. Even though KAN and YAB genes may have common targets, they also have different ones, since the phenotype of the fil yab3 double mutant is not quite the same as the extreme phenotype of kan1 kan2.

It has been hypothesized that the "default" state of cells is the abaxial fate. Genes that belong to the PHAB family (class III homeodomain-leucine zipper, HD-ZIP III;) of transcription factors, like PHB and PHV, might be activated by a proximal signal coming from the apical meristem. These cells that are programmed to yield the adaxial portion of the lateral organ, are predicted to in turn hava YAB and KAN genes repressed. In this respect, semidominant gain-of-function mutants of PHB and PHV genes cause adaxialization of lateral organs. PHB, PHV, and REV have similar expression patterns. They are expressed in the SAM initiating lateral organs and later become adaxially restricted as the primordium emerges. Finally, phenotypes of the loss-of-function rev mutants could be interpreted as having a partial loss of adaxial identity.

Besides the PHAB function in polarity, it is also interesting to note that a phb phv cna (corona, another member of the HDZIP III gene family) triple mutant has a very similar phenotype to those of clv mutants with a distinct increase in organ number in each whorl. This would suggest that HD-ZIP III genes and the CLV pathway regulate meristem function in a similar manner. The possible interrelation of these modules could contribute to homeostasis between stem cell maintenance and organ formation.

NUB and JAG are similar genes which encode C2H2-zinc finger transcription factors that are proposed to play redundant functions in proliferation and differentiation of adaxial cells, particularly during anther and carpel development. They specifically work together in determining the number of cell layers formed in floral organs, and like the PHAB family, they are not cell-fate genes. Hypothetically, JAG suppresses the premature differentiation of tissues by slowing down the cessation of cell division in distal regions of organs until it finally arrests after normal morphogenesis has occurred.

AS1 and AS2 have redundant functions in the establishment of adaxial identity. AS1 encodes a MYB-domain transcription factor, and AS2 is a member of the LBD gene family. AS1 protein is expressed in organ initials and physically interacts with AS2 to inhibit KNOX gene expression, thus guiding primordia toward differentiation.

Sepals and Petals

Sepals and petals constitute the sterile perianth in the first and second flower whorls, respectively. The sepal whorl or calyx protects the developing floral bud and in some plants, but not in Arabidopsis, it may be involved in fruit development. The petal whorl or corolla is generally thought to be important for attracting pollinators, but in an autogamous plant such as Arabidopsis, the corolla is generally not showy.

According to the ABC model, sepal identity specification depends on the activity of both A and *SEP* genes, and petal identity specification depends on the overlapping activities of A, B and *SEP* genes. Also, it has been shown that sepal and petal identity specification depends, at least in part, on the correct downregulation of *AG* expression in the second whorl.

Several molecular components known to influence development of sepals, influence petals too. But knowledge is still limited especially of sepal developmental gene networks. However, a basic GRN for petal development can be constructed based on available data. Organ identity determination, boundary establishment, and expression of polarity determinants are common features needed for the correct development of all the flower organs. There are several pieces of evidence that suggest that genes involved in these processes might be acting at the same time (for example, expression profiles and in situ hybridization assays), at least momentarily during flower development.

The sepal and petal boundary and organ number establishment are controlled by the CUC and FFO2 genes. CUC gene expression is regulated by the miR164c (encoded by EEP1) in an organ specific manner.

Several genes are involved in establishing and maintaining the sepal and/or petal domain and, in a way, determining the boundaries between the organs. One of the main activities of these genes is to exclude AG expression from the first and second whorl. AG is repressed by RBE, LUG, SEU, ROXY1, AP2, BLR, ANT and SAP.

Briefly, RBE is mainly involved in boundary and organ number determination of both sepals (non-autonomously) and petals, and in AG exclusion from the second whorl at early stages of flower organ development. But it is also important during late petal development as mutants may form filamentous organs in the second whorl. RBE expression is controlled by both PTL and UFO. PTL is a trihelix transcription factor that is expressed at early stages in four zones between the initiating sepal primordia and in lateral regions of stamen primordial. Later on, PTL expression can be detected at the margins of expanding sepals, petals, and stamens. Thus PTL may delimit the AG expression region indirectly by activating RBE expression, and it may also be controlling lateral outgrowth of mature sepals, petals and stamens defining their final shape and orientation.

UFO is also an important regulator of petal development. Its action toward RBE may be indirect, as it may be degrading (as part of an SCF E3 ubiquitin ligase complex) an unknown repressor of RBE. But UFO is an important network link between the AG inactivation pathway and the B gene identity determination pathway, because UFO interacts with LFY to activate AP3 expression. Importantly, UFO expression is also required for normal petal blade outgrowth after petal identity has been established, as well as for determination of sepal shape and number in the first whorl.

SEU and LUG also repress AG expression in the first and second whorls by forming a protein complex with AP1 and SEP3. But these genes are also part of the adaxial/abaxial polarity establishment pathway in the petal GRN, as they are required for normal PHB and FIL expression. SEU and LUG participate in petal shape regulation by controlling blade cell number and petal vasculature development in an AG independent manner. Finally, SEU is also involved in auxin response pathways by directly interacting with ETT, and influencing the final shape, number and phyllotaxy of petals.

As part of the regulatory network that represses AG expression, AP2 is itself negatively regulated by miR172 when second whorl boundaries are determined. Besides being a negative regulator of AG, ANT also affects organ number and morphology in the first three whorls. SAP, another regulator of the morphology of all organs, but mostly of petals, is unexpectedly more important in later flowers.

Another important indirect repressor of AG is ROXY1. As a glutaredoxin, ROXY1 seems to be a postranslational modifier of AP2, LUG, UFO and RBE giving them the specificity to repress AG in the second whorl. ROXY1 is also important for repressing PAN expression and for activating other TGA factors at different stages of petal development.

Genes that usually work in the establishment of lateral organ polarity are also important in determining the polarity of sepals and petals, e. g. PHB, JAG, FIL, YAB3, KAN, AS1 and AS2. Experimental data suggest that AS1, AS2 and JAG are negative regulators of CUC1/2 and PLT. This links the expression of these genes with those important for boundary determination in the GRN of both sepals and petals. PHB and FIL expression are also part of the network and are regulated by SEU and LUG. Lateralaxis dependent development is determined by the PRESSED FLOWER (PRS) homeobox gene. As with some other genes involved, its position in the GRN is unknown, but by analyzing the mutant phenotypes, it becomes clear that the same regulatory modules that underlie polarity determination are involved in organ shape regulation.

In Arabidopsis, as in other plants, several mutants featuring a foliose-sepal-syndrome (FSS) (leaf-like sepals) have been isolated. Ectopic expression of the MADS-box genes AGL24, SVP, and ZMM19 (from Zea mays), belonging to the STMADS11-clade, result in FSS. The main feature of these leaf-like sepals is that they are large and have leaf-like stellate trichomes on their outer surface. One of the characteristics of ap1 mutant plants is that they also have large or foliose sepals. Thus, it has been proposed that, in addition to their roles in floral transition and/or organ determination, AP1, SVP, and AGL24 may also have a role regulating sepal size. But how they interact among themselves or with other sepal specific genes is still unknown.

Final sepal and/or petal morphology is also determined by FRL1, the AP3/PI regulated genes GNC, GNL, At4g30270, HXK1, and NAP. Except for FRL1, which is involved in endoreduplication control, and TSO1, which is likely involved in chromatin remodelling, the position of these genes in the petal GRN has already been established.

Using microarray approaches Wellmer et al., compared gene expression levels within different floral homeotic mutants. Their first study of stage 2 flowers identified only 13 genes as being sepal-specific and only 18 genes expressed exclusively (or predominantly) in petals. However, a more recent study of flowers at stage 3, when sepal primordia have just formed, revealed that 199 genes are upregulated and 161 genes downregulated. One speculation is that sepals are relatively simple organs and not many specific genes are involved in their development. But more detailed studies are still required. Results also suggest that genes regulating sepal and petal development may have been recruited from leaf developmental pathways, and, hence, are not specific for the development of these organs.

Petals have been proposed as an excellent model system in which to study development because they have a simple organization and are not essential for survival or reproduction. Although much progress has been made, much work is still needed for an integrated and dynamical understanding of petal development.

Stamens

Six stamens occupy the third whorl in the Arabidopsis flower. Stamen specification depends on the overlapping activities of B, C and SEP MADS-box genes. A complex network of gene regulatory modules is simultaneously activated in young stamen primordia, and these are also important for organ morphogenesis. These modules include those that regulate adaxial-abaxial primordium polarity (also affecting other vegetative and reproductive lateral organs) including genes from the PHAB (PHB, PHV, and REV), KANADI (KAN1-4), and YABBY (FIUYAB1, YAB2, and YAB3) families. At later stages of stamen development, genes involved in sporogenesis such as SPL and BAM1/2, and in anther development, such as JAG and NUB, are activated.

Among the most striking stamen development mutants is fil (also called antherless and undeveloped anther) which bears normal filaments with neither anthers nor pollen. The FIL gene is YABBY-like and the fil phenotype suggests that the developmental programs of the filament and anther are controlled by independent regulatory modules.

SPL/NZZ is essential for male and female reproductive development and is probably the first reproductive gene to be activated in the anther or, at least, it is the only gene that remains active during most of early anther development. This transcription factor gene is expressed during micro- and megasporogenesis. AG directly induces SPL but AG is not necessary for maintaining its expression. spl mutants are not able to produce microsporogenous cells or tapetal tissue, and show several alterations in anther wall and nucellus development. Interestingly, BAM1 and BAM2, which participate in the first cell division of the archesporial cells and the subsequent periclinal divisions to produce the somatic cell layers, are proposed to form a regulatory loop with SPL. Since SPL maintains the sporogenous activity in the microsporogenous cells, and BAM1/2 maintain somatic differentiation, bam 1 bam2 anthers have cells interior to the epidermis with characteristics of pollen mother cells.

Although SPL is one of the genes expressed the earliest in stamen development, it is not the only one. Ectopic expression of SPL in all the whorls of an ag mutant, results in the formation of microsporangia only in the lateral parts of the staminoid 'petals', suggesting that microsporangial localization is established independently of AG, and that there is at least one other SPL inducer that is expressed in the second whorl, and not in other whorls. Two other genes, JAG and NUB play a crucial role in the formation of the four-locular anther architecture, independent of SPL induction, jag nub double mutants do not have a proper microsporangium. Instead, they form a finger-like structure that expresses SPL in its tips.

The correct number of microsporangial initials and the subsequent production of the tapetal cell and middle cell layer identities are properties specified by a putative LRR receptor kinase, EXCESS MICROSPOROCYTES1 (EMS1)/EXTRA SPOROGENOUS CELLS (EXS). Until recently, the ligand for EMS1 was unknown, though it was hypothesized that it could be involved in the same signaling pathway as the TAPETAL DETERMINANT (TPD1) gene. Both tpd1 and tpd1 ems1 mutants are similar to the single ems1 mutant with arrested meiotic cytokinesis and degenerated microsporocytes. TPD1, is a small putatively secreted protein that interacts with EMS1 and induces its phosphorylation suggesting that TPDI is the ligand of the EMS1 receptor that signals cell fate determination during sexual cell morphogenesis.

ROXY1 and ROXY2 redundantly regulate anther development in Arabidopsis. Lateral and medial stamens of roxy1 mutants might be fused and the former are sometimes missing. In these mutants, the adaxial anther lobes are affected in sporogenous cell formation during early differentiation steps, abaxial lobes develop normally but pollen mother cells degenerate, while the tapetum overgrows and occupies most of the locule space. Eventually, the tapetum degenerates too. RoXY1 and ROXY2 function downstream of SPL and upstream of DYSFUNCTIONAL TAPETUM1 (DVT1). As with other glutaredoxins, they may need an interaction with glutathione to catalyze biosynthetic reactions, suggesting that they may have a role in redox regulation and/or plant stress defense mechanisms involved in the control of male gametogenesis.

After tapetal cells are specified, a range of genes are essential for subsequent development. DYT1 encodes a putative bHLH transcription factor which functions downstream of SPL and EMS1. However DYT1 is not able to complement the spl or ems1 mutant phenotypes when it is overexpressed, indicating that it is required but not sufficient for normal tapetum development, dyt1 exhibits abnormal anther morphology with largely vacuolated tapetal cells that eventually collapse. Several tapetum-expressed genes, such as MALE STERILE 1 (MS1) and ABORTED MICRO-SPORES (AMS) are upregulated by DYT1. In ms1 mutants for example, tapetal cell abnormalities can be seen and pollen development is arrested just after microspores are released from the tetrads. Other genes that participate in tapetum development include RECEPTOR-LIKE PROTEIN KINASE2 (RPK2), FAT TAPETUM and GUS-NEGATIVE1 and 2 (GNE1, GNE2). RPK2 regulates tapetal function and middle layer differentiation. FAT TAPETUM, when mutated, has a middle layer that fails to collapse after meiosis and shows tapetal-like behavior. In gne1 and gne2 mutants the sporogenous cells enter meiosis, but cytokinesis is frequently arrested. The few highly aberrant tetrads formed degenerate early and microsporangia of mature anthers end up empty.

Several mutants affecting pollen development have been described: pollenless3; three division mutant (tdm1); ms5, ms3 and ms15; determinate infertile1 (dif1); switch1 (swi1); defectivepollen 1, and 6492 among others. Meiotic cells in pollenless3 anthers undergo a third division without DNA replication generating some cells with unbalanced chromosome numbers or "tetrads" with more than four microspores, dif1 and swi1 mutants have micro- and megaspores of uneven sizes because the encoded proteins are essential for sister chromatid cohesion in male and female meiosis and so mutants are totally infertile. Finally, other pollen mutants exhibit abnormal callose deposition (ms32, ms31, ms37, 7219, and 7593).

There are late-developmental anther mutants that affect anther dehiscence. In non-dehiscence1 mutant plants, anthers contain apparently wild-type pollen but do not dehisce. It has been hypothesized that a cell death suppression program, which is normally responsible for dehiscence, might be inactive in this mutant. ms35 is also affected in anther dehiscence, because endothecial cells fail to develop the lignified secondary walls that after desiccation shrink differentially leading to the retraction of the anther wall and full opening of the stomium. MS35, now MYB26, is expressed during early anther development and may be a regulator of NAC SECONDARY WALL-PROMOTING FACTOR 1 and 2 (NST1, NST2), which have also been linked to secondary thickening in the anther endothecium. In delayed-dehiscence mutants (dd1, dd2, dd3, dd4, dd5) anther dehiscence and pollen release occurs after the stigma is no longer receptive preventing successful pollination, but stamens look wild-type and pollen is viable. On the contrary, in defective-pollen1, 2, and 3, anthers are able to dehisce, but the pollen is aberrant and unviable.

The gibberellic acid (GA), jasmonic acid (JA), and auxins are involved during stamen development. The GA-deficient mutant, *ga1-3*, produces an abortive anther where microsporogenesis is arrested prior to pollen mitosis. Mutations in two GA receptors, GA-INSENSITIVE DWARF1a and b (AtGID1a, b), affect the elongation of stamens, suggesting that these receptors have specific roles during stamen development. GA induces the degradation of the DELLA protein REPRESSOR OF GA1–3 (RGA) upon ubiquitination. Microarray analysis shows that 38% of the RGA downregulated genes are expressed in the male gametophyte at various stages of microsporogenesis.

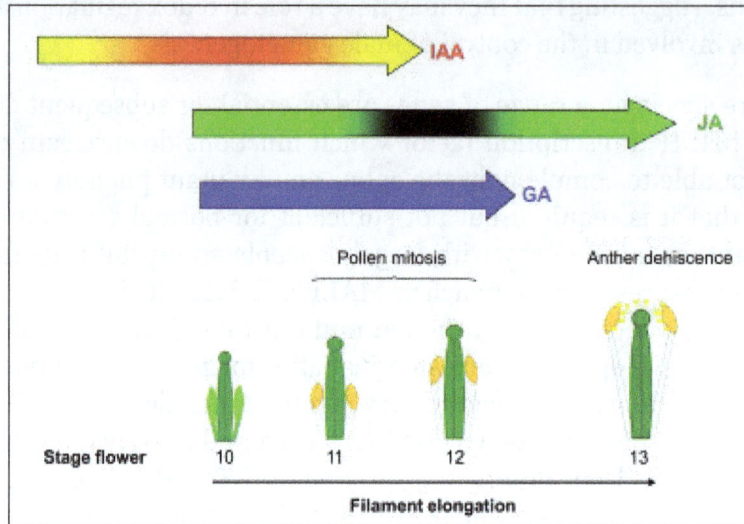

Hormones in late stages of stamen development.

At stage 10 of flower development, the auxin (IAA) concentration (yellow arrow) peaks (red gradient) in the stamens. During this period filaments start to elongate and auxin prevents premature dehiscence. Auxin also participates in later anther dehiscence, probably by inducing JA production (green arrow) that peaks (dark green gradient) at stages 11 and 12. JA coordinates filament elongation, pollen maturation, anther dehiscence and flower opening. Although it has not been quantified, GA (blue arrow) is involved in filament elongation and participates in microsporogenesis. Pollen development in anthers of GA-biosynthetic mutants is arrested before microspore mitosis.

Temporal coordination of the elongation of filaments, pollen maturation, and dehiscence of anthers is important for efficient fertilization. The expression overlap of RGA-regulated genes and jasmonate-responsive genes during stamen development suggest a crosstalk between GA and JA signaling pathways in these processes.

JA has been shown to be involved in at least three androecial developmental pathways: filament elongation, anther dehiscence and pollen production. Different male sterile mutants have been found to be JA biosynthetic mutants including: the triple fed mutant (fad3–2 fad7–2 fad8), which lacks the fatty acid precursors of JA; defective in anther dehiscence 1 (dad1), which encodes a phospholipase A1 that catalyzes the initial step of JA biosynthesis; and dd1, a member of the 12-OXO-PHYTODIENOATE REDUCTASE (OPR3) gene family. OPR3/DD1 is expressed in the stomium and in the septum cells of the anther that are involved in pollen release. All these mutant phenotypes can be rescued by exogenous application of JA, suggesting that this hormone plays an important role in controlling the timing of anther dehiscence. Interestingly, DAD1 is a direct target of AG.

Similarly, the coronatine insensitive 1 (coi1; JA receptor) mutant is defective in both pollen development and anther dehiscence. Stamens of coi1 flowers have shorter filaments than those of wild-type flowers and anthers are indehiscent containing pollen grains with conspicuous vacuoles.

Three related polygalacturonases, enzymes involved in pectin degradation that promotes cell separation, are also involved in JA-regulated anther dehiscence. ARABIDOPSIS DEHISCENCE ZONE POLYGALACTURONASE 1 (ADPGI) and 2 (ADPG2), and QUARTET2 (QRT2) gene expression are distinctly regulated by JA.

To determine the jasmonate-regulated stamen-specific transcriptome the expression profiles of JA-treated and untreated opr3 mutants were compared. It was found that 821 genes were induced (70% of them expressed in the sporophytic tissue) and 480 genes were repressed by JA and 13 transcription factors were identified that could be important for stamen maturation pathways. Of these, MYB21, MYB24, and MYB28 are JA-responsive genes. myb21 mutants have short filaments, are late to dehisce and have reduced fertility. Though myb24 mutants look like wild type, myb21 myb24 double mutants have a more severe phenotype than myb21, suggesting that these two genes might be redundantly involved in important aspects of JA-dependent stamen development. MYB28 is involved in amino acid metabolism and it is downregulated by both JA and RGA. This also uncovered several other biochemical pathways that could be important during stamen and pollen maturation. Other results indicate that JA coordinates pollen maturation, anther dehiscence, and flower opening. Auxins have also been proved to participate in these processes arf6 arf8 double mutants are defective in anther dehiscence probably because they produce too little JA. Accordingly, this phenotype can be rescued by application of JA. However, auxins trigger filament elongation and prevent premature anther dehiscence and pollen maturation at earlier stages of stamen development. While JA production peaks at stages 11–12 of flower development auxin receptors (TIR1 and AFBs) are already expressed at the end of meiosis. Mutants in these genes cause the release of mature pollen grains before filaments elongate. At later stages, the amount of JA decreases allowing these processes to continue.

Additional stamen or pollen microarray analyses have been performed recently. For example, a clear difference was found between the genes that are expressed in the sporophyte and in pollen with 39% of the expressed genes being pollen specific. The global gene expression profiles of wild-type reproductive axes have been compared to those of the floral mutants ap3, spl/nzz, and ms1 in order to study gene expression during stamen development and microspore formation. The data suggest that different interconnected regulatory modules may control specific stages of anther and microspore development.

Carpels and Ovules

Carpels are specified by the C gene AG, and the SHP1, SHP2, and STK genes (in an AG independent manner) together with the SEP genes. They arise in the center of the flower meristem and when carpels are fully developed the floral meristematic cells are completely consumed. Carpels are the most complex structures within flowers and a GRN underlies their development.

Three different stages of carpel development are represented by the schemes in the upper part of the figure. Briefly, at stage 6, the central zone of the FM begins to grow upward and eventually will form the gynoecium. From stages 11 to 13, the ovule primordia (O) arise from the placenta flanking the medial ridges, and the Archesporial cell (Ar) develops from a single hypodermal cell at the ovule. The Ar then forms the megaspore mother cell (MMC) through megasporogenesis, and the MMC forms the embryo sac through megagametogenesis. The embryo sac consists of 2 synergids, 1 egg cell, 1 central cell and 3 antipodal cells. The medial ridges meet in the center of the fruit to form the septum (sm) which divides the gynoecium in two internal compartments. The mature gynoecium is externally formed by the fusion of two valves (va); internally, it also carries totally differentiated ovules each one containing its own embryo sac.

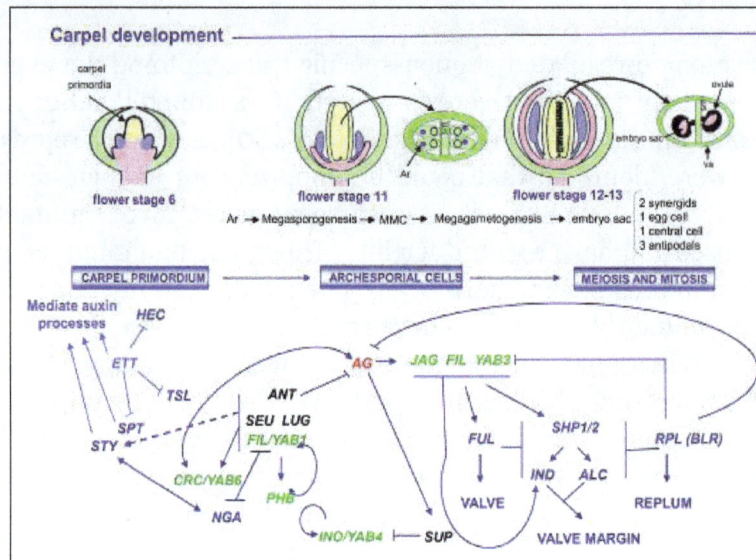

Main stages of carpel development and some genes involved.

Carpel-specific gene networks are shown in blue. Part of the network shown here was taken from Roeder and Yanofsky and Balanzá et al. Color codes of interactions and gene/floral organs are according to those of functional modules identified in figure. Arrows and bars indicate positive and negative regulatory interactions, respectively.

Nectaries

Little is known about the molecular genetics of nectary development. It is clear that nectaries are ABC-independent because ap2-2 pi-1 ag-1 triple mutant flowers develop nectaries, although in these mutants nectaries are clearly smaller. However, ABC genes may play a role in nectary patterning as pi-1 ag-1 and ap3-3 ag-3 double mutants lack them. Also, single mutant lfy and ufo plants show reduced nectary formation.

Several genes have been found to be expressed in the nectaries (e. g., SHP1, ALC, SPL, MS35 and XAL1), but no detectable defect is observed in their respective mutants. The only gene that has been clearly related to nectary development is CRC, which is also important for gynoecial development. The regulation of CRC in the nectaries seems to be independent of its expression in the gynoecium. Expression of this gene is already detectable at stage 6 of flower development in the area where the nectaries will be formed. Thus, CRC may be important for the early specification of nectary cells. CRC may also be necessary for maturation or maintenance of the nectaries, because it is expressed at high levels when they develop and crc mutants have defects in nectary development. But CRC is not sufficient for nectary development, because its ectopic expression does not yield ectopic nectaries. Bioinformatic and functional molecular genetic approaches have been taken to discover components of the regulatory network in which CRC participates during nectary and carpel development. A combination of floral homeotic gene activities acting redundantly with each other, involving AP3, PI and, AG and in combination with SEP proteins, activate CRC in both organs. The CRC was also found to be a direct target gene of AG and to be indirectly regulated by LFY and UFO. A model has been proposed in which LFY and UFO activate downstream MADS-box genes which could be working in conjunction with region-specific factors to activate CRC during nectary and carpel development.

Evolutionary studies have indicated that the CRC gene may have been recruited as a regulator of nectary development in the core eudicot plant lineage, but its ancestral function could have been related to carpel development.

References

- Ronse De Craene, L. P. (2007). "Are Petals Sterile Stamens or Bracts? The Origin and Evolution of Petals in the Core Eudicots". Annals of Botany. 100 (3): 621–630. Doi:10.1093/aob/mcm076. PMC 2533615. PMID 17513305

- Flowerstructure, plants-human: plantphys.info, Retrieved 31 August, 2019

- Stearn, William Thomas (2004). Botanical Latin (p/b ed.). David & Charles/Timber Press. ISBN 978-0-7153-1643-6. P. 39

- Carr, Gerald. "Lythraceae". University of Hawaii. Archived from the original on 2008-12-05. Retrieved 2008-12-20

- Soltis, Pamela S.; Douglas E. Soltis (2004). "The origin and diversification of angiosperms". American Journal of Botany. 91 (10): 1614–1626. Doi:10.3732/ajb.91.10.1614. PMID 21652312

- Kessler, Danny; Kallenbach, Mario; Diezel, Celia; Rothe, Eva; Murdock, Mark; Baldwin, Ian T (2015-07-01). "Abstract". Elife. 4. Doi:10.7554/elife.07641.001. ISSN 2050-084X

4
Ornamental Plants and their Cultivars

The plants which are grown for display purposes instead of functional purposes are known as ornamental plants. They include various kinds of cultivars such as rose cultivars, orchid cultivars, banksia cultivars, grevillea cultivars and callistemon cultivars. This chapter has been carefully written to provide an easy understanding of ornamental plants and these cultivars.

Ornamental Plant

Plants grow all over the world in different sizes, shapes and appearance. Some provides us with food, shelter or building materials, while others provide us with only visual delight. Ornamental Plants are also referred to as garden plants has beauty as its main trait. They are usually grown in the flower garden for the display of their flowers. It is a plant primarily grown for its beauty either for screening, accent, specimen, color or aesthetic reasons. Common ornamental features include leaves, scent, fruit, stem and bark.

The history of ornamental gardening started at least 4, 000 years of human civilization. Egyptian tomb paintings of the 1500 BC are some of the earliest physical evidence of ornamental horticulture and landscape design. It depicts depict lotus ponds surrounded by symmetrical rows of acacias and palms.

Examples of Ornamentals

Daffodil

Lilium Stagazer lily.

Rocky-Mountain Iris.

Ornamental plants are grown for use by the green industry and public for purposes such as landscaping for sport, and conservation. The green industries include commercial plant nurseries, flower growers, parks, and roadside and landscape plant installation and maintenance.

The primary use for these plants is not for food, fuel, fiber, or medicine. However, ornamental plants contribute significantly to the quality of life by acting as barriers to wind, providing cooling shade, reducing or eliminating erosion, cleaning the air and water of pollutants including dust and chemicals, reducing noise pollution, and providing food and habitat for wildlife while making both suburban and urban areas more beautiful. Their economic and emotional impact is significant.

Ornamental plants include perennial deciduous and evergreen shade trees, conifers, and shrubs grown in horticultural production by the commercial nursery industry. Ornamental plants also include herbaceous and woody indoor and outdoor landscape broadleaf plants, grasses, and palms produced by traditional floricultural and nursery techniques within greenhouses, shaded structures, and other environments significantly modified to favor healthy, rapid, and profitable plant growth.

Ornamentals include annual, biennial, or perennial plants. They may be field grown in native or amended soils and then harvested and marketed with native soils intact. This form of horticulture is generally referred to as "balled and burlapped" (B&B) plant production even though burlap may not be used in their harvest. They may also be harvested without soil and referred to as "bare root." The most popular method of growing ornamental plants is in soilless growing media within containers. Soilless growing media are most often natural organic materials such as peat or tree bark mixed with a mineral component such as sand or perlite.

Cultivars

A cultivar is a plant or group of plants that have been selected from a naturally occurring species and bred to enhance or maintain a particular set of desirable characteristics. These plants almost always originate from human cultivation, propagated through cutting or grafting, and often cannot be grown from seeds from the original plant.

The term cultivar was created by the botanist Liberty Hyde Bailey as a combination of the words cultivated and variety. Although the term variety is sometimes used interchangeably with cultivar, a variety is different because it can be obtained from a seed and will maintain the characteristics of the original plant. However, a seed from the original plant of a cultivar may be very different and maintain no similar characteristics.

Most cultivars are food crops or ornamental plants. Many ornamental plants, such as roses and azaleas, are cultivated to enhance a particular flower shape, size, or color. In edible plants, desirable characteristics include abundant yield, pleasant taste, and resistance to disease.

Orchid Cultivars

Phalaenopsis Kaleidoscope

Phalaenopsis Baldan's Kaleidoscope is an artificial orchid hybrid. One of the seedlings from this grex was awarded an Award of Merit by the American Orchid Society and given the cultivar name 'Golden Treasure'. The vast majority of Phal. Baldan's Kaleidoscope plants sold today are clones of this original plant and is one of the most cloned orchid in the world. Widely sold in stores, used by florist and found in most orchid collections. It has many showy yellow flowers with red to deep pink stripes on a well formed arching spike. It is well known not just to be an attractive orchid, but it is hardy and blooms well.

Phalaenopsis Baldan's Kaleidoscope 'Golden Treasure' has been used successfully as the seed parent of a few new hybrids, for example, *Phal.* Baldan's Geisha = *Phal.* Baldan's Kaleidoscope × *Phal.* Golden Buddha.

Papilionanthe Miss Joaquim

Papilionanthe Miss Joaquim also known as Vanda Miss Joaquim, the Singapore orchid and the Princess Aloha orchid and incorrectly as Vanda 'Miss Agnes Joaquim', is a hybrid orchid cultivar that is Singapore's national flower. For its resilience and year-round blooming quality, it was chosen in 1981 to represent Singapore's uniqueness and hybrid culture.

Features

It is a free flowering plant and each inflorescence can bear up to 12 buds, and usually 4 flower blossom at a time. Each flower measures 5 cm across and 6 cm tall. The petals are twisted such that the back surface faces the front like its parents. The two petals on the top and the top sepal are rosy-violet, while the 2 lateral sepals on the lower half are pale mauve. The large and board lip of the orchid which looks like a fan is colored violet-rose, and merges into a contrasting fiery orange that are finely spotted with dark purple center. Papilionanthe 'Miss Joaquim' is a robust sun loving plant that requires heavy fertilizing, vertical support to enable it to grow straight and tall along with free air movement and high humidity. It starts blossoming after its stem rises 40 to 50 cm above the support.

Kimilsungia

Kimilsungia is a hybrid orchid of the genus Dendrobium. It is a clone of a plant that was created in Indonesia by orchid breeder Carl Ludwig C. L. Bundt, who in 1964 registered the grex name Dendrobium Clara Bundt for all orchids of the same ancestry, naming it after his daughter. It has a complex ancestry from cultivated orchids. An attempt was made to register the grex name Dendrobium Kimilsungia, but this is not valid, it is a later synonym of Dendrobium Clara Bundt. As a cultivar name (applying to only part of the grex), the correct name would be Dendrobium Clara Bundt 'Kimilsungia'. Another grex name Dendrobium Kimilsung Flower refers to plants of related but different ancestry.

Another flower, the Kimjongilia, is named after Kim Il-sung's son, Kim Jong-il. Neither the Kimilsungia nor the Kimjongilia are the national flower of North Korea. The national flower of the country is the Magnolia sieboldii with white flowers. The Kimilsungia violet orchid has become an integral part of the ever-present state-sponsored propaganda that surrounds the late leader.

According to the Korean Central News Agency, Kim Il-sung's "peerless character" is "fully reflected in the immortal flower" which is "blooming everywhere on the five continents".

According to the Pyongyang-published book Korea in the 20th Century: 100 Significant Events, Kim Il-sung travelled to Indonesia to meet with his counterpart, Sukarno. Kim was taken on a tour of the Bogor Botanical Garden, where:

> "He stopped before a particular flower, its stem stretching straight, its leaves spreading fair, giving a cool appearance, and its pink blossoms showing off their elegance and preciousness; he said the plant looked lovely, speaking highly of the success in raising it. Sukarno said that the plant had not yet been named, and that he would name it after Kim Il Sung. Kim Il Sung declined his offer, but Sukarno insisted earnestly that respected Kim Il Sung was entitled to such a great honour, for he had already performed great exploits for the benefit of mankind".

The plant grows 30 to 70 centimetres (12–28 in) high. Its leaves adhere to the nodes alternatively and each stalk yields 3-15 flowers. The flowers have three petals and three calyxes and measure 6 to 8 centimetres (2.4–3.1 in). It blooms for 60–90 days. It grows best in daylight temperatures of 25 to 30 °C (77 to 86 °F) and 18 to 23 °C (64 to 73 °F) at night.

Annual Festivals

The annual Kimilsungia Festival has been held since 1998, and is held around the Day of the Sun. Kimilsungia flower shows are held every year in Pyongyang. Traditionally, embassies of foreign countries in North Korea each present their own bouquet of the flower to the annual exhibition.

Rose Cultivars

Garden Roses

Garden roses are predominantly hybrid roses that are grown as ornamental plants in private or public gardens. They are one of the most popular and widely cultivated groups of flowering plants, especially in temperate climates. Numerous cultivars have been produced, especially over the last two centuries, though roses have been known in the garden for millennia beforehand. While most garden roses are grown for their flowers, some are also valued for other reasons, such as having ornamental fruit, providing ground cover, or for hedging.

The Hybrid Tea rose, 'Peer Gynt'.

Features

An amber-coloured rose.

Roses are one of the most popular garden shrubs in the world with both indoor and outdoor appeal. They possess a number of general features that cause growers and gardeners to choose roses for their gardens. This includes the wide range of colours they are available in; the generally large size of flower, larger than most flowers in temperate regions; the variety of size and shape; the wide variety of species and cultivars that freely hybridize.

Colour of Flowers

Rose flowers have historically been cultivated in a diverse number of colours ranging in intensity and hue; they are also available in countless combinations of colours which result in multicoloured flowers. Breeders have been able to widen this range through all the options available with the range of pigments in the species. This gives us yellow, orange, pink, red, white and many combinations of these colours. However, they lack the blue pigment that would give a true purple or blue colour and until the 21st century all true blue flowers were created using some form of dye. Now, however, genetic modification is introducing the blue pigment. Colours are bred through plant breeding programs which have existed for a long time. Roses are often bred for new and intriguing colour combinations which can fetch premium prices in market.

Classification

There is no single system of classification for garden roses. In general, however, roses are placed in one of three main groups: Wild, Old Garden, and Modern Garden roses. The latter two groups are usually subdivided further according to hybrid lineage, although due to the complex ancestry of most rose hybrids, such distinctions can be imprecise. Growth habit and floral form are also used as means of classification. This is the most common method to classify roses as it reflects their growth habits.

Wild Roses

The spring-flowering pimpinellifolia 'Rosa Altaica', underplanted with lamium.

Wild roses, also denominated "species roses", include the natural species and some of their immediate hybrid descendants. The wild roses commonly grown in gardens include Rosa moschata ("musk rose"), Rosa banksiae ("Lady Banks' rose"), Rosa pimpinellifolia ("Scots rose" or "burnet rose"), Rosa rubiginosa ("sweetbriar" or "eglantine"), and Rosa foetida in varieties 'Austrian Copper', 'Persian Double', and 'Harison's Yellow'. For most of these, the plants found in cultivation are often selected clones that are propagated vegetatively. Wild roses are low-maintenance shrubs in comparison to other garden roses, and they usually tolerate poor soil and some shade. They generally have only one flush of blooms per year, described as being "non-remontant", unlike remontant, modern roses. Some species have colorful hips in autumn, e. g. Rosa moyesii, or have colourful autumnal foliage, e. g. Rosa virginiana.

Old Garden Roses

An old garden rose is defined as any rose belonging to a class which existed before the introduction of the first modern rose, La France, in 1867. Alternative terms for this group include heritage and historic roses. In general, Old Garden roses of European or Mediterranean origin are once-blooming woody shrubs, with notably fragrant, double-flowered blooms primarily in shades of white, pink and crimson-red. The shrubs' foliage tends to be highly disease-resistant, and they generally bloom only from canes (stems) which formed in previous years. The introduction of China and Tea roses from East Asia around 1800 led to new classes of Old Garden Roses which bloom on new growth, often repeatedly from spring to fall. Most Old Garden Roses are classified into one of the following groups.

Alba

'Maiden's Blush', an Alba rose.

Literally "white roses", derived from R. arvensis and the closely allied R. × alba. The latter species is a hybrid of R. gallica and R. canina. This group contains some of the oldest garden roses. The shrubs flower once yearly in the spring or early summer with scented blossoms of white or pale pink. They frequently have gray-green foliage and a vigorous or climbing habit of growth. Examples are 'Alba Semiplena', 'White Rose of York'.

Gallica

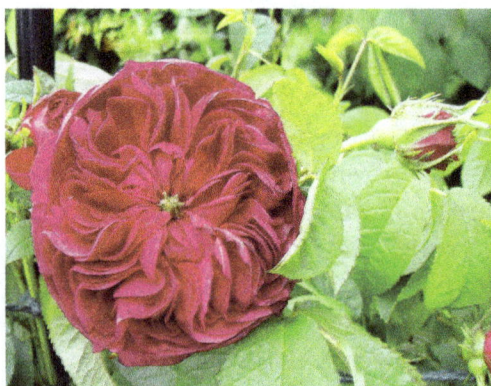

Gallica rose 'Charles de Mills', ante.

The Gallica, Gallica Hybrid, or Rose of Provins group is a very old class developed from Rosa gallica, which is a native of central and southern Europe and western Asia. The "Apothecary's Rose", R. gallica varietas officinalis, was grown in the Middle Ages in monastic herbaria for its alleged medicinal properties, and became famous in English history as the Red Rose of Lancaster. Gallicas are shrubs that rarely grow over 4 feet (1.25 m) tall and flower once in Summer. Unlike most other once-blooming Old Garden Roses, gallicas include cultivars with flowers in hues of red, maroon, and purplish crimson. Examples include 'Cardinal de Richelieu', 'Charles de Mills', and 'Rosa Mundi' (R. gallica varietas versicolor).

Damask

'Autumn Damask' ('Quatre Saisons').

Named for Damascus, Damask roses (Rosa × damascena) originated in ancient times with a natural hybrid (Rosa moschata × Rosa gallica) × Rosa fedtschenkoana. Robert de Brie is given credit for bringing damask roses from the Middle East to Europe sometime between 1254 and 1276, although there is evidence from ancient Roman frescoes that at least one damask rose existed in Europe for hundreds of years before this. Summer damasks bloom once in summer. Autumn or Four Seasons damasks bloom again later, albeit less exuberantly, and these were the first remontant (repeat-flowering) Old European roses. Damask roses tend to have rangy to sprawling growth habits and strongly scented blooms. Examples: 'Ispahan', 'Madame Hardy'.

Centifolia or Provence

Centifolia roses are also known as Cabbage roses, or as Provence roses. They are derived from *Rosa × centifolia*, a hybrid that appeared in the 17th century in the Netherlands, related to damask roses. They are named for their "one hundred" petals; they are often called "cabbage" roses due to the globular shape of the flowers. The centifolias are all once-flowering. As a class, they are notable for their inclination to produce mutations of various sizes and forms, including moss roses and some of the first miniature roses. Examples: 'Centifolia', 'Paul Ricault'.

Moss

The Moss roses are based on one or more mutations, particularly one that appeared early on Rosa × centifolia, the Provence or cabbage rose. Some with Damask roses as a parent may be derived from a separate mutation. Thickly growing or branched resin-bearing hairs, particularly on the sepals, are considered to resemble moss and give off a pleasant woods or balsam scent when rubbed. Moss roses are cherished for this trait, but as a group they have not contributed to the development of new rose classifications. Various hybrids with other roses have yielded different forms, such as the modern miniature creeping moss rose 'Red Moss Rambler'. Moss roses with centifolia background are once-flowering; some moss roses exhibit repeat-blooming, indicative of Autumn

Damask parentage. Examples: 'Common Moss' (centifolia-moss), 'Mousseline', also known as 'Alfred de Dalmas' (Autumn Damask moss).

Portland

The Portland roses were long thought to be the first group of crosses between China roses and European roses, and to show the influence of *Rosa chinensis*. Recent DNA analysis however has demonstrated that the original Portland Rose has no Chinese ancestry, but has an autumn damask/gallica lineage. This group of roses was named after the Duchess of Portland who received (from Italy about 1775) a rose then known as *R. paestana* or 'Scarlet Four Seasons' Rose' (now known simply as 'The Portland Rose'). The whole class of Portland roses was developed from that one rose. The first repeat-flowering class of rose with fancy European-style blossoms, the plants tend to be fairly short and shrubby, with a suckering habit, with proportionately short flower stalks. The main flowering is in the summer, but intermittent flowers continue into the autumn. Examples: 'James Veitch', 'Rose de Rescht', 'Comte de Chambord'.

China

'Parson's Pink China' or 'Old Blush, ' one of the "stud Chinas".

The China roses, based on *Rosa chinensis*, have been cultivated in East Asia for centuries. They have been cultivated in Western Europe since the late 18th century. They contribute much to the parentage of today's hybrid roses, and they brought a change to the form of the flowers then cultivated in Europe. Compared with the older rose classes known in Europe, the Chinese roses had less fragrant, smaller blooms carried over twiggier, more cold-sensitive shrubs. However they could bloom repeatedly throughout the summer and into late autumn, unlike their European counterparts. The flowers of China roses were also notable for their tendency to "suntan, " or darken over time unlike other blooms which tended to fade after opening. This made them highly desirable for hybridisation purposes in the early 19th century. According to Graham Stuart Thomas, China roses are the class upon which modern roses are built. Today's exhibition rose owes its form to the China genes, and the China roses also brought slender buds which unfurl when opening. Tradition holds that four "stud China" roses—'Slater's Crimson China' (1792), 'Parsons' Pink China' (1793), and the Tea roses 'Hume's Blush Tea-scented China' (1809) and 'Parks' Yellow Tea-scented China' (1824)—were brought to Europe in the late 18th and early 19th centuries; in fact there were rather more, at least five Chinas not counting the Teas having been imported. This brought about the creation of the first classes of repeat-flowering Old Garden Roses, and later the Modern Garden Roses. Examples: 'Old Blush China', 'Mutabilis' (Butterfly Rose), 'Cramoisi Superieur'.

Tea

Tea rose 'Mrs Dudley Cross'.

The original Tea-scented Chinas (Rosa × odorata) were Oriental cultivars thought to represent hybrids of R. chinensis with R. gigantea, a large Asian climbing rose with pale-yellow blossoms. Immediately upon their introduction in the early 19th-century breeders went to work with them, especially in France, crossing them first with China roses and then with Bourbons and Noisettes. The Tea roses are repeat-flowering roses, named for their fragrance being reminiscent of Chinese black tea (although this is not always the case). The colour range includes pastel shades of white, pink and (a novelty at the time) yellow to apricot. The individual flowers of many cultivars are semi-pendent and nodding, due to weak flower stalks. In a "typical" Tea, pointed buds produce high-centred blooms which unfurl in a spiral fashion, and the petals tend to roll back at the edges, producing a petal with a pointed tip; the Teas are thus the originators of today's "classic" florists' rose form. According to rose historian Brent Dickerson, the Tea classification owes as much to marketing as to botany; 19th-century nurserymen would label their Asian-based cultivars as "Teas" if they possessed the desirable Tea flower form, and "Chinas" if they did not. Like the Chinas, the Teas are not hardy in colder climates. Examples: 'Lady Hillingdon', 'Maman Cochet', 'Duchesse de Brabant', 'Mrs. Foley Hobbs'.

Bourbons

Bourbon rose Rosa 'Souvenir de la Malmaison'.

Bourbon roses originated on the Île Bourbon (now called Réunion) off the coast of Madagascar in the Indian Ocean. They are believed to be the result of a cross between the Autumn Damask and the 'Old Blush' China rose, both of which were frequently used as hedging materials on the island. They flower repeatedly on vigorous, frequently semi-climbing shrubs with glossy foliage and purple-tinted canes. They were first Introduced in France in 1820 by Henri Antoine Jacques. Examples: 'Louise Odier', 'Mme. Pierre Oger', 'Zéphirine Drouhin' (the last example is often classified under climbing roses).

Noisette

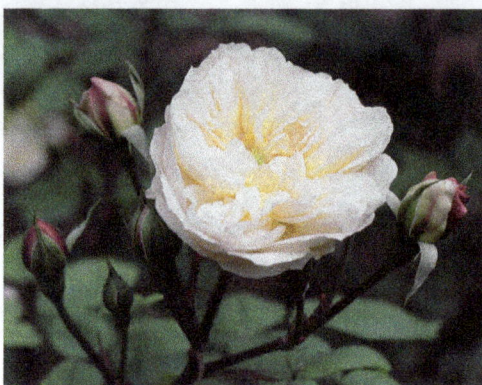

Noisette rose 'Desprez à fleurs jaunes'.

The first Noisette rose was raised as a hybrid seedling by a South Carolina rice planter named John Champneys. Its parents were the China rose 'Parson's Pink' and the autumn-flowering musk rose (Rosa moschata), resulting in a vigorous climbing rose producing huge clusters of small pink flowers from spring to fall. Champneys sent seedlings of his rose (called 'Champneys' Pink Cluster') to his gardening friend, Philippe Noisette, who in turn sent plants to his brother Louis in Paris, who then introduced 'Blush Noisette' in 1817. The first Noisettes were small-blossomed, fairly winter-hardy climbers, but later infusions of Tea rose genes created a Tea-Noisette subclass with larger flowers, smaller clusters, and considerably reduced winter hardiness. Examples: 'Blush Noisette', 'Lamarque' (Noisette); 'Mme. Alfred Carriere', 'Marechal Niel' (Tea-Noisette).

Hybrid Perpetual

Hybrid perpetual rose 'La Reine'.

The dominant class of roses in Victorian England, hybrid perpetuals, their name being a misleading translation of "hybrides remontants" ("reblooming hybrids"), emerged in 1838 as the first roses which successfully combined Asian remontancy (repeat blooming) with the old European lineages. Because remontancy is a recessive trait, the first generation of Asian and European crosses, i. e., hybrid Chinas, hybrid bourbons, and hybrid noisettes, were stubbornly non-remontant, but when these roses were re-crossed with themselves or with Chinas or teas, some of their offspring flowered more than once. The hybrid perpetuals thus were something of a miscellaneous, catch-all class derived to a great extent from the bourbons but with admixtures of Chinas, teas, damasks, gallicas, and to a lesser extent noisettes, albas, and even centifolias. They became the most popular garden and florist roses of northern Europe at the time, as the tender tea roses would not thrive in cold climates, and the hybrid perpetuals' very large blooms were well suited to the new phenomenon of competitive exhibitions. The "perpetual" in the name suggests their remontancy, but many varieties of this class only poorly exhibited the property; the tendency was for a massive vernal bloom followed by either scattered summer flowering, a smaller autumnal burst, or sometimes no re-flowering in that year. Due to a limited colour palette of white, pink, and red, and a lack of reliable remontancy, the hybrid perpetuals were finally overshadowed by their descendants, the hybrid teas. Examples include 'Général Jacqueminot', 'Ferdinand Pichard', 'Paul Neyron', and 'Reine des Violettes'.

Hybrid Musk

Hybrid musk rose 'Moonlight'.

Although they arose too late to qualify technically as old garden roses, the hybrid musks are often informally classed with them, since their growth habits and care are much more like the old garden roses than modern roses. The hybrid musk group was mainly developed by Rev. Joseph Pemberton, a British rosarian, in the first decades of the 20th century, based upon 'Aglaia', an 1896 cross by Peter Lambert. A seedling of this rose, 'Trier', is considered to the foundation of the class. The genetics of the class are somewhat obscure, as some of the parents are unknown. Rosa multiflora, however, is known to be one parent, and Rosa moschata (the musk rose) also figures in its heritage, though it is considered to be less important than the name would suggest. Hybrid musks are disease-resistant, repeat flowering and generally cluster-flowered, with a strong, characteristic "musk" scent. The stems tend to be lax and arching, with limited thorns. Examples include 'Buff Beauty' and 'Penelope'.

Hybrid Rugosa

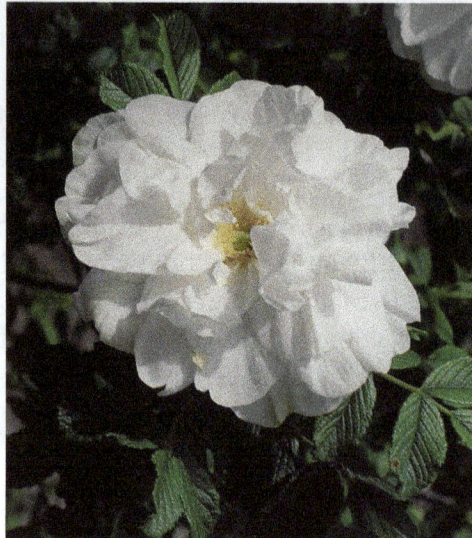

Rugosa rose 'Blanc Double de Coubert'.

The hybrid rugosas likewise are not officially old garden roses, but tend to be included in them. Derived from *Rosa rugosa* ("Japanese rose"), native to Japan and Korea and introduced into the West circa the 1880s, these vigorous roses are extremely hardy with excellent disease resistance. Most are extremely fragrant, remontant, and produce moderately double, flat flowers. The defining characteristic of a hybrid rugosa rose is its rugose, i. e., wrinkly foliage, but some hybrid rugosas lack this trait. These roses often set large hips. Examples include 'Hansa' and 'Roseraie de l'Häy'.

Bermuda "Mystery" Roses

This is a group of several dozen "found" roses grown in Bermuda for at least a century. The roses have significant value and interest for those growing roses in tropical and semi-tropical regions, since they are highly resistant to both nematode damage and the fungal diseases that plague rose culture in hot, humid areas. Most of these roses are thought to be Old Garden Rose cultivars that have otherwise dropped out of cultivation, or sports thereof. They are "mystery roses" because their "proper" historical names have been lost. Tradition dictates that they are named after the owner of the garden where they were rediscovered.

There are also a few smaller classes (such as Scots, Sweet Brier) and some climbing classes of old roses (including Ayrshire, Climbing China, Laevigata, Sempervirens, Boursault, Climbing Tea, and Climbing Bourbon). Those classes with both climbing and shrub forms are often grouped together.

Modern Garden Roses

Classification of modern roses can be quite confusing because many modern roses have old garden roses in their ancestry and their form varies so much. The classifications tend to be by growth and flowering characteristics. The following includes the most notable and popular classifications of Modern Garden Roses:

Hybrid Tea

A 'Memoriam' hybrid tea rose.

The favourite rose for much of the history of modern roses, hybrid teas were initially created by hybridising hybrid perpetuals with tea roses in the late 19th century. 'La France', created in 1867, is universally acknowledged as the first indication of a new class of roses. Hybrid teas exhibit traits midway between both parents: hardier than the teas but less hardy than the hybrid perpetuals, and more ever-blooming than the hybrid perpetuals but less so than the teas. The flowers are well-formed with large, high-centred buds, and each flowering stem typically terminates in a single shapely bloom. The shrubs tend to be stiffly upright and sparsely foliaged, which today is often seen as a liability because it makes them more difficult to place in the garden or landscape. Hybrid teas became the single most popular garden rose of the 20th century; today, their reputation as high maintenance plants has led to a decline in popularity. The hybrid tea remains the standard rose of the floral industry, however, and is still favoured in formal situations. Examples: 'Peace' (yellow), 'Garden Party' (white), 'Mister Lincoln' (red) and 'Double Delight' (bi-colour cream and red).

Pernetiana

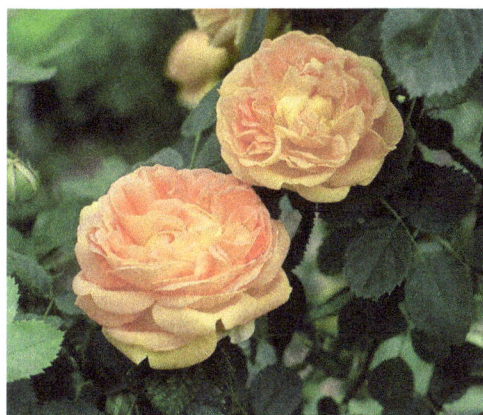

Pernetiana rose 'Soleil d'Or', the first of its class.

The French breeder Joseph Pernet-Ducher initiated the first class of roses to include genes from the old Austrian briar rose (*Rosa foetida*) with his 1900 introduction of 'Soleil d'Or. ' This resulted in an entirely new colour range for roses: shades of deep yellow, apricot, copper, orange, true scarlet, yellow bicolours, lavender, gray, and even brown were now possible. Originally considered a separate class, the Pernetianas or Hybrid Foetidas were officially merged into the Hybrid Teas in 1930. The new colour range did much to increase hybrid tea popularity in the 20th century, but these colours came at a price: *Rosa foetida* also passed on a tendency toward disease-susceptibility, scentless blooms, and an intolerance of pruning to its descendants.

Polyantha

Originally derived from crosses between two East Asian species, Rosa chinensis and Rosa multiflora, polyanthas first appeared in France in the late 19th century alongside the hybrid teas. They are short plants, some compact and others spreading in habit, producing tiny blooms (2.5 cm or 1 inch in diameter on average) in large sprays in the typical rose colours of white, pink, and red. Their popularity derived from their prolific blooming: from spring to autumn; a healthy polyantha shrub truly might be covered in flowers, making a strong colour impact in the landscape.

Polyantha roses are still popular and regarded as low-maintenance, disease-resistant, garden roses. The class of roses denominated "Multiflora Hybrids" are probably cognizable as polyanthas. Examples include Rosa 'Cécile Brünner', 'The Fairy', 'Pink Fairy', and 'Red Fairy'.

Floribunda

Rosa 'Borussia', a modern floribunda rose.

Some rose breeders recognized potential in crossing polyanthas with hybrid teas, to create roses that bloomed with the profusion of polyanthas and possessed the floral beauty and breadth of coloration of hybrid teas. In 1907, the Danish breeder Dines Poulsen introduced the first cross of a polyantha and hybrid tea, denominated 'Rödhätte'. This hybrid had some characteristics of both of its parental classes, and was denominated a "Hybrid Polyantha" or "Poulsen" rose. Further and similar introductions followed from Poulsen, these often bearing the family name, e. g., 'Else Poulsen'. Because their hybrid characteristics separated them from polyanthas and hybrid teas, the new class eventually was named *Floribunda*. Typical floribundas are stiff shrubs that are smaller and bushier than the average hybrid tea, but less dense and sprawling than the average polyantha. The flowers are often smaller than those of hybrid teas but are produced in large sprays that give a better floral effect in the garden. Floribundas are found in all hybrid tea colours and often with the classic, hybrid tea-shaped blossom. Sometimes they differ from hybrid teas only in their cluster flowered habit. They are still planted in large bedding schemes in public parks and suitable gardens. Examples include 'Anne Harkness', 'Dainty Maid', 'Iceberg', and 'Tuscan Sun'.

Grandiflora

Grandifloras are the class of roses created in the middle of the 20th century as back crosses of hybrid teas and floribundas that fit neither category, specifically, Rosa 'Queen Elizabeth', which was introduced in 1954. Grandiflora roses are shrubs that are typically larger than both hybrid teas and floribundas and produce flowers that resemble those of hybrid teas and are borne in small clusters of three to five, similar to floribundas. Grandifloras were somewhat popular from circa 1954 into the 1980s, but today they are much less popular than both hybrid teas and floribundas. Examples include: 'Comanche, ' 'Montezuma', and 'Queen Elizabeth'.

Miniature

'Meillandine' (a miniature rose) in a terracotta flowerpot.

Dwarf mutations of some Old Garden Roses—gallicas and centifolias—were known in Europe in the 17th century, although these were once-flowering just as their larger forms were. Miniature forms of repeat-flowering China roses were also grown and bred in China, and are depicted in 18th-century Chinese art. Modern miniature roses largely derive from such miniature China roses, especially the cultivar 'Roulettii', a chance discovery found in a pot in Switzerland.

Miniature roses are represented by twiggy, repeat-flowering shrubs ranging from 6" to 36" in height, with most falling in the 12"–24" height range. Blooms come in all the hybrid tea colours; many varieties also emulate the classic high centred hybrid tea flower shape. Miniature roses are often marketed and sold by the floral industry as houseplants, but it is important to remember that these plants are largely descended from outdoor shrubs native to temperate regions; thus, most miniature rose varieties require an annual period of cold dormancy to survive. (Examples: 'Petite de Hollande' (Miniature Centifolia, once-blooming), 'Cupcake' (Modern Miniature, repeat-blooming). Additional examples: Scentsational, Tropical Twist. Miniature garden roses only grow in the summer.

Climbing and Rambling

Rosa 'Zéphirine Drouhin', a climbing Bourbon rose.

The classes of roses, both Old and Modern, have "climbing/arching" forms, whereby the canes of the shrubs grow to be much longer and more flexible than the normal "bush" forms. In the Old Garden Roses, this is often simply the natural growth habit; for many Modern Roses, however, climbing roses are the results of spontaneous mutations. For example, 'Climbing Peace' is designated as a "Climbing Hybrid Tea, " for it is genetically identical to the normal "shrub" form of the 'Peace' hybrid tea rose, except that its canes are long and flexible, i. e. "climbing. " Most Climbing Roses grow 6–20 feet tall and exhibit repeat blooming.

The "Peggy Martin Rose" survived 20 feet of salt water over
the garden of Mrs. Peggy Martin, Plaquemines Parish, Louisiana,
after Hurricane Katrina. It's a thornless climbing rose.

"Rambler Roses", although technically a separate class, are often included in Climbing Roses. They also exhibit long, flexible canes, but are usually distinguished from true climbers in two ways: a larger overall size (20–30 feet tall is common) and of a once-blooming habit. Climbing and

Rambling Roses are not true vines such as ivy, clematis, and wisteria because they lack the ability to cling to supports on their own and must be manually trained and tied over structures, such as arbors and pergolas. Examples include 'American Pillar' (once-blooming rambler), and 'Blaze' (repeat-blooming climber).

One of the most vigorous of the Climbing Roses is the Kiftsgate Rose, *Rosa filipes* 'Kiftsgate', named after the house garden where Graham Stuart Thomas noticed it in 1951. The original plant is claimed to be the largest rose in the United Kingdom, and has climbed 50 feet high into a copper beech tree.

Shrub

The shrub rose 'Mollineux'.

This is not a precisely defined class of garden rose, but it is a description or grouping commonly used by rose reference books and catalogues. It encompasses some old single and repeat flowering cultivars, as well as modern roses that don't fit neatly into other categories. Many cultivars placed in other categories are simultaneously placed in this one. Roses classed as shrubs tend to be robust and of informal habit, making them recommended for use in a mixed shrub border or as hedging.

English / David Austin

Austin rose 'Abraham Darby'.

Although not officially recognized as a separate class of roses by any established rose authority, English roses are often set aside as such by consumers and retailers alike. Development started in the 1960s by David Austin of Shropshire, England, who wanted to rekindle interest in Old Garden Roses by hybridizing them with modern hybrid teas and floribundas. The idea was to create a new group of roses that featured blooms with old-fashioned shapes and fragrances, evocative of classic gallica, alba and "damask" roses, but with modern repeat-blooming characteristics and the larger modern colour range as well. Austin mostly succeeded in his mission; his tribe of "English" roses, now numbering hundreds of varieties, has been warmly embraced by the gardening public and are widely available to consumers. David Austin roses are still actively developed, with new varieties released regularly. The typical winter-hardiness and disease-resistance of the classic Old Garden Roses has largely been compromised in the process; many English roses are susceptible to the same disease problems that plague modern hybrid teas and floribundas, and many are not hardy north of USDA Zone 5. Examples: 'Charles Austin', 'Graham Thomas', 'Mary Rose', 'Tamora', 'Wife of Bath'.

Canadian Hardy

Rosa 'Henry Hudson', one of the Explorer series.

Two main lines of roses were developed for the extreme weather conditions of Canadian winters by Agriculture Canada at the Morden Research Station in Morden, Manitoba and the Experimental Farm in Ottawa (and later at L'Assomption, Québec). They are called the Explorer series and the Parkland series. These programs have now been discontinued, the remaining plant stock has been taken over by private breeders and marketed along with the Canadian Artists roses as a single series. Derived mostly from crosses of Rosa rugosa or the native Canadian species Rosa arkansana with other species, these plants are extremely tolerant of cold weather, some down to −35 °C. All have repeat bloom. A wide diversity of forms and colours were achieved.

'Thérèse Bugnet', a multi-species hybrid
that is still widely available.

Examples of roses in the Explorer series are: 'Martin Frobisher', 'Jens Munk' (1974), 'Henry Hudson' (1976), 'John Cabot' (1978), 'David Thompson' (1979), 'John Franklin' (1980), 'Champlain' (1982), 'Charles Albanel' (1982), 'William Baffin' (1983), 'Henry Kelsey' (1984), 'Alexander Mackenzie' (1985), 'John Davis' (1986), 'J. P. Connell' (1987), 'Captain Samuel Holland' (1992), 'Frontenac' (1992), 'Louis Jolliet' (1992), 'Simon Fraser' (1992), 'George Vancouver' (1994), 'William Booth' (1999).

Roses in the Parkland series include 'Morden Centennial', 'Morden Sunrise, 'Winnipeg Parks' and 'Cuthbert Grant'. Two roses named after Canadian artists that have been added are 'Emily Carr' and 'Felix Leclerc'.

Landscape (Ground Cover)

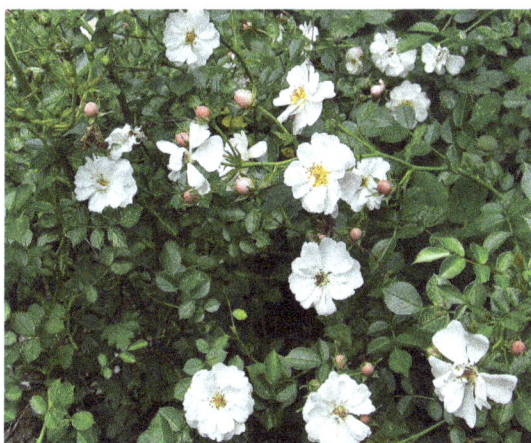

'Avon', a ground cover rose introduced by Poulson in 1992.

This type of rose was developed mainly for mass amenity planting. In the late 20th century, traditional hybrid tea and floribunda rose varieties fell out of favour with many gardeners and landscapers, as they are often labour and chemical intensive plants susceptible to pest and disease problems. So-called "landscape" roses (also known as "ground cover" roses) have thus been developed to fill the consumer desire for a garden rose that offers colour, form and fragrance, but is also low maintenance and easy to care for. Most have the following characteristics:

- Lower growing habit, usually under 60 cm (24 inches).

- Repeat flowering.

- Disease and pest resistance.

- Growing on their own roots.

- Minimal pruning requirements.

Patio

Since the 1970s many rose breeders have focused on developing compact roses (typically 1'–4' in height and spread) that are suitable for smaller gardens, terraces and containers. These combine characteristics of larger miniature roses and smaller floribundas—resulting in the rather loose classification "patio roses", called Minifloras in North America. D. G. Hessayon says the description "patio roses" emerged after 1996. Some rose catalogues include older polyanthas that have stood the test of time (e. g., 'Nathalie Nypels', 'Baby Faurax') within their patio selection. Rose breeders, notably Chris Warner in the UK and the Danish firm of Poulson (under the name of Courtyard Climbers) have also created patio climbers, small rambler style plants that flower top-to-toe and are suitable for confined areas. It is suggested patio style roses are protected during winter months due to the exposure environment.

Chris Warner's patio climber 'Open Arms'.

Cultivation

In the garden, roses are grown as bushes, shrubs or climbers. "Bushes" are usually comparatively low growing, often quite upright in habit, with multiple stems emerging near ground level; they are often grown formally in beds with other roses. "Shrubs" are usually larger and have a more informal or arching habit, and may additionally be placed in a mixed border or grown separately as specimens. Certain bush hybrids (and smaller shrubs) may also be grown as "standards", which are plants grafted high (typically 1 metre or more) on a rose rootstock, resulting in extra height which can make a dominant feature in a floral display. Climbing roses are usually trained to a suitable support.

Roses with protection against freezing.

Roses are commonly propagated by grafting onto a rootstock, which provides sturdiness and vigour, or (especially with Old Garden Roses) they may be propagated from hardwood cuttings and allowed to develop their own roots.

Most roses thrive in temperate climates. Those based on warm climate Asian species do well in their native sub-tropical environments. Certain species and cultivars can even flourish in tropical climates, especially when grafted onto appropriate rootstocks. Most garden roses prefer rich soil which is well-watered but well-drained, and perform best in well-lit positions which receive several hours of sun a day (although some climbers, some species and most Hybrid Musks will tolerate shade). Standard roses require staking.

Pruning

Rose pruning, sometimes regarded as a horticultural art form, is largely dependent on the type of rose to be pruned, the reason for pruning, and the time of year it is at the time of the desired pruning.

Most Old Garden Roses of strict European heritage (albas, damasks, gallicas, etc.) are shrubs that bloom once yearly, in late spring or early summer, on two-year-old (or older) canes. Their pruning requirements are quite minimal because removal of branches will remove next year's flower buds. Hence pruning is usually restricted to just removing weak and spent branches, plus light trimming (if necessary) to reduce overall size.

Modern hybrids, including the hybrid teas, floribundas, grandifloras, modern miniatures, and English roses, have a complex genetic background that almost always includes China roses (which are descended from *Rosa chinensis*). China roses were evergrowing, everblooming roses from humid subtropical regions that bloomed constantly on any new vegetative growth produced during the growing season. Their modern hybrid descendants exhibit similar habits; unlike Old European Roses, modern hybrids bloom continuously (until stopped by frost) on any new canes produced during the growing season. They therefore require pruning back of any spent flowering stem in order to divert the plant's energy into producing new growth and hence new flowers.

Additionally, modern hybrids planted in cold winter climates will almost universally require a "hard" annual pruning (reducing all canes to 8"–12", about 30 cm in height) in early spring. Again, because of their complex China rose background, modern hybrids are typically not as cold hardy as European Old Garden Roses, and low winter temperatures often desiccate or kill exposed canes. In spring, if left unpruned, these damaged canes will often die back all the way to the shrub's root zone, resulting in a weakened, disfigured plant. The annual "hard" pruning of hybrid teas and floribundas is generally done in early spring.

Deadheading

This is the practice of removing any spent, faded, withered, or discoloured flowers. The purpose is to encourage the plant to focus its energy and resources on forming new shoots and blooms, rather than fruit production. Deadheading may also be performed for aesthetic purposes, if spent flowers are unsightly. Any roses such as *Rosa glauca* or *Rosa moyesii* that are grown for their decorative hips should not be deadheaded.

Pests and Diseases

Roses are subject to several diseases. The main fungal diseases affecting the leaves are rose black spot (Diplocarpon rosae), rose rust (Phragmidium mucronatum), rose powdery mildew (Sphaerotheca pannosa) and rose downy mildew (Peronospora sparsa). Stems can be affected by several canker diseases, the most commonly seen of which is stem canker (Leptosphaeria coniothyrium). Diseases of the root zone include honey fungus (Armillaria spp.), verticillium wilt, and various species of phytophthora.

Fungal leaf diseases affect some cultivars and species more than others. On susceptible plants fungicidal sprays may be necessary to prevent infection or reduce severity of attacks. Cultivation techniques may also be used, such as ensuring good air circulation around a plant. Stem cankers are best treated by pruning out infection as soon as it is noticed. Root diseases are not usually possible to treat once infection has occurred; the most practical line of defence is to ensure that growing conditions maximise plant health and thereby prevent infection. Phytophthora species are waterborne and therefore improving drainage and reducing waterlogging can help reduce infection.

The main pest affecting roses is the aphid (greenfly), which sucks the sap and weakens the plant. In areas where they are endemic Japanese beetles (Popillia japonica) take a heavy toll on rose flowers and foliage; rose blooms can also be destroyed by infestations of thrips (Thysanoptera spp). Roses are also used as food plants by the larvae of some Lepidoptera (butterfly and moth) species. Spraying with insecticide of roses is often recommended but if this is done care is needed to minimize the loss of beneficial insects; systemic insecticides have the advantage of only affecting insects which feed on the plants.

Rosa 'Double Delight'

Rosa 'Double Delight', is a multiple award winning, red blend hybrid tea rose cultivar bred in the United States by Swim & Ellis and introduced in 1977. Its parents were two hybrid tea cultivars, the red and yellow 'Granada' and the ivory 'Garden Party'.

The large, strongly fragrant red and white flowers have a high-centered bloom form and appear continuously throughout the season. They are double, have a diameter of more than 10 cm (4") and up to 30 petals. In the sun, their colour changes from white to carmine red, beginning at the edges. The flowers are very large and usually borne singly, on long, prickly stems. The flowers have a strong, sweet scent. The red color of 'Double Delight' is the product of ultraviolet light on natural pigments in the petals. If the plant is grown in a greenhouse, which blocks ultraviolet light, the petals will be white in color.

'Double Delight' is a medium-tall bushy shrub with large, dark green leaves. It grows about 90 to 150 cm (3' to 5') high, and 60 centimetres (2.0 ft) wide. It is winter hardy up to −25 °C (USDA zone 5), but can be susceptible to mildew and black spot. The cultivar needs sunny, warm places, but can be grown in containers. 'Double Delight' is used as garden rose and as cut flower.

Sports

One sport of 'Double Delight' was introduced in 1982. The climbing sport, Rosa 'Double Delight, Cl', also known as 'AROclidd' and 'Grimpant Double Delight' was discovered by John Nieuwesteeg and Jack E. Christensen. The cultivar is a popular rose and similar to 'Double Delight', except that it climbs quickly to great heights. The plant, when mature, will bloom in abundance once a year, and will continue to flower, but in smaller numbers.

Rosa 'Constance Spry'

Rosa 'Constance Spry' was the first rose bred by David C. H. Austin to be released, in 1961. Developed at a time when the hybrid tea rose was the most usual style marketed to gardeners, it renewed the popularity of a more old fashioned type of rose. The rose is named after educator and florist Constance Spry.

A cross between the pink floribunda, Dainty Maid, and the *gallica* rose, Belle Isis, the fully double flowers are cupped and globular in shape; a pale pink on the outside, and a deeper, darker pink within. Growing between 8.5 to 20 feet (larger in warmer areas) and described as a "lanky grower" it can be grown as a climber or large shrub. Constance Spry flowers only once a year and is noted for its strong, distinctive "myrhh like" scent, a characteristic inherited by many of its descendants.

Despite its only once a year flowering, the charm of its old fashioned appearance proved popular enough to prove there was a market for "reproduction" style roses, and Austin continued his breeding program. Constance Spry was further crossed with both modern and older roses, resulting in the fully remontant Wife of Bath and Chaucer, from which many of his later roses descended.

Rosa 'Chrysler Imperial'

Rosa 'Chrysler Imperial' is a strongly fragrant, dark red hybrid tea rose cultivar. This variety was bred and publicly debuted by Dr. Walter E. Lammerts of Descanso Gardens, La Cañada Flintridge, California, US in 1952. Its stock parents 'Charlotte Armstrong' (cerise pink) and 'Mirandy' (dark oxblood red) are both 'All American Rose Selections'-roses (awarded in 1940 and 1945).

The elegantly tapered buds open into high-centered blossoms with a diameter of about 11–13 cm (4.5–5 in) and can have up to 45–50 petals (which is a high number for a hybrid tea rose) with a rich, deep, velvety red color. The cultivar flushes in a chronological blooming pattern throughout its local season, starting in late spring until fall. The long-stemmed rose flowers are long lasting and showy and make excellent cut flowers, though they "blue" badly with age.

The rose bush reaches 75 to 200 cm (30 to 79 in) height, and a diameter of 60 to 120 cm (24 to 47 in). The shrub has an upright form with very thorny canes and semi-glossy dark green foliage. It is not a very cold hardy rose and needs good sun exposure. Without good air circulation it is susceptible to mildew and blackspot, particularly in cool climates.

Rosa 'Charles Austin'

Rosa 'Charles Austin' (Ausfather) is an apricot rose cultivar bred and introduced by David Austin in England in 1973. The rose was hybridised by crossing the English rose 'Chaucer' with the pink Hybrid Tea 'Aloha' and is named after the breeder's father. It was one of the early English roses.

Double, flat or slightly cupped flowers with a strong, fruity fragrance, and an average diameter of 10 cm (4 inches) appear in small cluster of 3 to 5 in flushes throughout the season. Their colour is an apricot blend, with stronger colours at the petal base, fading to cream at the edges. The flowers have about 70 petals arranged in a quartered bloom form, with the outer ones lighter than the inner ones, and are well suited as cut flowers.

The tall and bushy shrub can grow well in excess of 200 cm, especially in warmer climates and is somewhat slow to rebloom, especially if not drastically pruned after the first flush. The cultivar has large leaves and fine, red prickles, is winter hardy up to –20 °C (USDA zone 5b – 10b), but susceptible to black spot and mildew. Due to its size, it can be grown as a freestanding shrub, pegged or trained as a small climber.

'Charles Austin' was further used by Austin as a parent rose and fathered the cultivars 'Leander' (1982), 'Graham Thomas' (1983), 'Swan' (1987), 'Brother Cadfael' (1990), 'Golden Celebration' (1992), 'Tradescant' (1993), 'Teasing Georgia' (1998) and 'Benjamin Britten' (2001). In 1981, Austin introduced a sport (mutation) – 'Yellow Charles Austin' – with lemon to golden yellow colours, that fade to cream.

Charles Austin (Ausfather) is one of a number of varieties which has been 'retired' by the David Austin Roses company in favour of other more modern and healthy varieties.

Banksia Cultivars

Banksia 'Giant Candles'

Banksia 'Giant Candles' is a registered Banksia cultivar. It is a hybrid between the Gosford form of B. ericifolia (heath-leaved banksia) and a form of B. spinulosa var. cunninghamii.

It looks like a shrub, and this form, that is equally broad as tall, can grow up to 5 metres. It is well known for its extremely large flower spikes, which easily can become 40 cm long. They have a habit of drooping or bending occasionally. The flowers are a bronzy-orange and will be showy from late autumn through winter. They grow in most well-drained soils, and will flower best if grown in full sun. In an ideal area, they will grow up to 800 mm per year. In areas where irrigation is limited, they will not produce a heavy canopy, but will produce about 30% shade under its evergreen foliage.

Russell Costin of Limpinwood Nursery, who originally propagated and registered it in the 1970s, has reported its popularity waned for a few years but has become more popular in the last decade. Angus Stewart reported it to be iron hungry, so treat yellowing with iron chelate or iron sulfate.

Banksia 'Yellow Wing' is a hybrid derived from Banksia Giant Candles and Banksia spinulosa var. collina.

Banksia 'Yellow Wing'

Banksia 'Yellow Wing' is a Banksia cultivar developed by Austraflora Nurseries of Dixons Creek in Victoria, Australia. The cultivar grown to about 1.8 metres in both height and width and has large gold inflorescences held above the foliage.

It is a hybrid between Banksia 'Giant Candles' and Banksia spinulosa var. collina from Carnarvon Gorge in Queensland. The culivar name references the yellow-winged honeyeater which is attracted to the inflorescences.

The cultivar is suitable to hedge or screen planting and flowers can be cut and used fresh or dried. It is tolerant of a range of climatic conditions and can withstand mild frosts. It prefers a position in full sun or partial shade, and is adaptable to dry conditions once established.

Banksia 'Superman'

Banksia 'Superman', also known by its extended cultivar name Banksia serrata 'Superman', is a registered Banksia cultivar. It was discovered by Maria Hitchcock of Armidale NSW near Nambucca in 1986 during the Banksia Atlas project. An attempt to have it accorded subspecies rank was not successful so she named it 'Superman' to describe the giant inflorescences and leaves and in keeping with the common name for Banksia serrata (Saw Banksia). Its leaves and inflorescences are mostly twice the size of typical plants of its parent species, Banksia serrata. Naturally occurring close to running water or on poorly drained sites between Nambucca Heads and Grassy Head in New South Wales, it grows true to seed.

It has not yet been introduced into commercial cultivation but seed has been distributed among members of the Australian Plants Society. Specimens have been growing successfully in the Armidale district for more than 15 years and in Canberra. The variety is frost hardy especially when it achieves some height but it is only moderately drought hardy. It has a short warty trunk and thick branches. The upright and terminal inflorescences which occur on short thickened stems off the branches are grey in bud and up to 25 cm x 12 cm in size. The yellow styles emerge in a spiral at the bottom of the inflorescence and gradually cover the whole inflorescence. Lorikeets and other birds are attracted to the nectar. The fruiting cone is covered with dead brown styles and has prominent follicles which contain one or two large seeds with black papery wings separated by a woody spacer. Cones need to be heated in a fire or oven for the follicles to open. The tree drops leaves continually creating a layer of mulch.

Grevillea Cultivars

Grevillea lanigera 'Mt Tamboritha'

Grevillea lanigera 'Mt Tamboritha' is a cultivar of the genus Grevillea, planted widely in Australia and other countries for its ornamental foliage and flowers. It is the most popular form of Grevillea lanigera in cultivation. It is also known by the names 'Mt Tamboritha form', 'Compacta', 'Prostrate', 'Prostrate Form' or the misnomer 'Mt Tambourine'.

The cultivar is a spreading, low shrub that grows to 0.4 metres in height and 1 to 2 metres wide. The overall appearance of the foliage is dense and dark green with a silvery sheen from the fine hairs. The leaves, which are narrow and oblong, are notably smaller than most forms of the species. These are arranged spirally on the arching, wiry stems. Flowers are pinkish-red and cream and appear in clusters. The primary flowering period is from late winter to spring, though flowers may be seen throughout the year.

The cultivar was introduced by plantsman Bill Cane. Despite being widely known as the "Mount Tamboritha form" of Grevillea lanigera it is believed that it did not originate from Mount

Tamboritha, a mountain in the Alpine National Park, east of Licola in Victoria. A form of Grevillea lanigera occurs at that location, but it is different in appearance. It is thought that the cultivar is a coastal form that may have been selected from the Yanakie Isthmus - Wilsons Promontory region in Victoria. 'Mt Tamboritha' has not been registered with the Australian Cultivar Registration Authority.

Cultivation

The cultivar has been planted widely in Australia and other countries due its ornamental foliage, compact form, showy flower clusters and attraction to nectar-seeking birds. Plants are suited to being grown in rockeries or containers.

'Mt Tamboritha' prefers a well-drained situation with full sun exposure, or in partial shade. The preferred soil type is sandy to medium loam that is slightly acidic and well-drained. It can grow in a wide variety of climates, being able to withstand temperatures as low as −5 °C (23 °F), while tolerant of humid subtropical climates such as in Brisbane. Once established, it will tolerate extended dry periods. In wet or humid situations, the foliage underneath may turn brown, a situation that may be overcome by using a stone or pebble mulch underneath.

In 2003, it was reported that the fungal disease Phytophthora palmivora had been detected in plant nurseries in Sicily, leading to root rot and death of potted Grevillea cultivars. However, plants of 'Mt Tamboritha' appeared resistant.

Propagation from cuttings is required to ensure new plants are true to type.

Grevillea 'Peaches and Cream'

Grevillea 'Peaches and Cream' is new and much sought-after grevillea cultivar which has been recently released in Australia.

It is a shrub that grows to 1.2 by 1.5 metres (4–5 ft) in height and width and has bright green attractive deeply divided leaves, around 11.7 centimetres (4.6 in) long by 6.4 centimetres (2.5 in) in width. The foliage takes on a bronze sheen in winter. The inflorescences are about 15 centimetres (5.9 in) long by 9 centimetres (3.5 in) wide and open yellow initially but later add various shades of pink and orange.

The cultivar is a cross between a white-flowered form of the Queensland species Grevillea banksii, and G. bipinnatifida from Western Australia, and was selected from a plant which arose in a garden in Logan Village, a southern suburb of Brisbane, in 1997. It was watched and propagated by Queensland horticulturists and SGAP members Dennis Cox and Janice Glazebrook, finally being patented in 2006.

It is of the same parentage as 'Superb' and 'Robyn Gordon' and has similar prolific and sustained flowering. Grevillea 'Superb' has a deeper orange coloration in the flowers, while G. 'Robyn Gordon' is red.

Its small size lends itself to use in a small garden, and it is bird attracting. It tolerant of a wide range of conditions, including humidity as well as drought, and frost down to −5 °C (23 °F).

Although not yet recorded, the cultivar is very similar to several cultivars which have been known to cause allergic contact dermatitis for certain individuals who come into contact with it, so caution is advised.

Grevillea 'Moonlight'

Grevillea 'Moonlight' is a widely cultivated and popular garden plant in Australian gardens and amenities. It was a selected form of the Queensland species Grevillea whiteana, although this has been questioned because of the difference in appearance to the parent plant. A hybrid between a white-flowered form of Grevillea banksii and the previous plant has been proposed.

It is an upright woody shrub which may reach 4 m (13 ft) high by 1.5 m (5 ft) wide. It has deeply divided dark green fern-like leaves that are approximately 20 cm (8 in) long. The inflorescences are creamy white racemes that are up to 25 cm (10 in) long, and may occur year-round.

Highly regarded by celebrity gardener Don Burke among others, it has been widely used in gardens and amenities plantings around Australia, where it thrives in a well-drained sunny position. It is tolerant of humidity and frost. As with all cultivars, propagation is by cuttings, though this can be difficult. Heavy pruning may be required to keep it from getting top-heavy as well as promoting a dense habit.

It has also been used in the cut flower industry to some extent, as well as proposed as a suitable plant for street plantings.

Callistemon Cultivars

Callistemon 'Splendens'

Callistemon 'Splendens' is a commonly grown cultivar of the plant genus Callistemon. It has a compact and rounded habit and usually grows to about 2 metres (6 ft 7 in) high and wide, although it may grow taller. Large, well-displayed "brushes" are produced in late spring, with further flowering sometimes occurring at other times. New growth is pink-tinged and the leaves are elliptic and up to 90 mm long and 20 mm wide.

The cultivar, which has been in grown for many years, is of uncertain origin. It was originally known as Callistemon citrinus var. splendens, first formally described in 1925 in Botanical Magazine. In 1970 it was promoted under the name 'Endeavour' to mark the bicentennial of James Cook's voyage to Australia on the Endeavour. It was registered under the name Callistemon 'Splendens' with the Australian Cultivar Registration Authority in 1989.

This plant has gained the Royal Horticultural Society's Award of Garden Merit.

Cultivation

The cultivar is most suited to climates ranging from temperate to sub-tropical. It adapts to most soils, and is tolerant of frost and salt spray.

Callistemon 'Lilacinus'

Callistemon 'Lilacinus' is a cultivar of the genus Callistemon. It grows to between 2.5 and 4 metres high and has purplish-violet inflorescences. Leaves are smooth and sharp pointed, with thick margins and are 40 to 100 mm long and 6 to 18 mm wide.

The cultivar was first selected in Berlin in 1913 from plants raised from seed collected near Como, New South Wales by German botanist Ernst Betsche in 1894. At this time, it was referred to as *C. lanceolatus* var *lilacina*. In 1925 Edwin Cheel gave the cultivar the species name *Callistemon lilacinus*, but this was later brought in line with other cultivars and amended to *C.* 'Lilacinus'.

The cultivar has often been referred to as *C.* 'Violaceus'.

Two other forms of *C. lilacinus* were described by Cheel:

- C lilacinus f. albus, which was "discovered in nature" at Long Bay by Mr H. Burrell, and currently given the name C. citrinus 'Lilacinus Albus'.

- C lilacinus f. carminus, a reddish-purple flowering form exhibited at the Linnean Society of New South Wales and currently given the name C. citrinus 'Lilacinus Carminus'. It was raised by E. Ashby in South Australia.

Uses of Ornamental Plants

An ornamental plant is grown for decoration, rather than food or other by-products. Ornamental plants may be grown in a flowerbed, shaped into a hedge or placed in a sunny apartment window. They are most often intentionally planted for aesthetic appeal, but a plant that occurs naturally and enhances the landscape could also be considered ornamental. While the most apparent use of ornamental plants is for visual effect, they serve a few less obvious purposes.

Add Beauty

Ornamental plants are used in landscapes and throughout the home to beautify the surroundings. A large, tropical plant in a living room provides a pop of color and helps soften harsh lines from furniture and architectural design. Colorful flowering ornamental plants break up the browns and greens that naturally occur outside. A large dogwood tree in the center of the front yard awakens with brilliant pink or white blossoms to flood the yard with color in spring. Exposed concrete block foundations are commonly concealed with hedges of boxwood, privet and other shrubs. Even fruit and vegetable trees and plants are sometimes used ornamentally when the plants lend themselves in some way to improving the visual appeal of the landscape.

Fragrance

Many ornamental plants are chosen because they appeal to the sense of smell, in addition to their visual appeal. Lavender is widely regarded for its pleasing fragrance; although widely harvested for lavender oil, it is commonly planted in home landscapes for its scent while in bloom. Roses are another type of flower well known for their pleasing scent. A walk through a rose garden is sure to entice visitors to lean in for a whiff of the floral bouquet.

Some fragrant plants prove beneficial at repelling outdoor pests, such as ants, mosquitoes and flies. Perhaps the most well known is the citronella plant, a type of geranium with a lemon fragrance. The fuzzy blossoms of ageratum plants are prevalent in flowerbeds, but the flowers also produce coumarin, a natural mosquito repellent.

Attract Wildlife

Ornamental plants provide nutrition and shelter for many wildlife species. While some forms of wildlife wreak havoc on carefully planned landscapes, other species are responsible for pollination and propagation, making this attraction vital to the ecosystem. Choosing native plant species ensures there are ornamental plants in your garden adapted to attract native wildlife. Fruit- and berry-producing plants attract birds and small animals. Ornamental plants that produce berries include hawthorn, crabapple and native plants such as baneberry and Pacific madrone. Entire gardens of ornamentals are often dedicated to attracting butterflies. Plants to include in a butterfly garden are purple coneflower, coast angelica, coast buckwheat and pipevine. Twinberry is a native species that provides food for insects and hummingbirds with its flowers, and its berries are food for other birds.

Clean Air

Without plants we wouldn't have clean air to breathe, because plants create oxygen during photosynthesis. Plants take in carbon dioxide as food and release clean oxygen, acting as natural air filters. This proves especially helpful for indoor environments, where air circulation is limited compared to outdoors. Keeping ornamental houseplants has been shown to improve indoor air quality, even removing tobacco smoke and such volatile organic compounds as formaldehyde, trichloroethylene and benzene. Plants that prove especially effective include spider plant, golden pothos, peace lily, snake plant and several species of philodendron and dracaena. The healthier a plant is, the more effective it is at removing harmful toxins from the air. A National Aeronautics and Space Administration study suggests using up to 18 plants in 6- to 8-inch containers to clean the air in an 1, 800-square-foot house.

References

- Liber, C. (2003). "Update on Eastern Banksia Cultivars" (PDF). Banksia Study Group Newsletter. 5: 1–5. Archived from the original (PDF) on 14 April 2008. Retrieved 3 August 2008

- Ornamental-plants, news-wires-white-papers-and-books, science: encyclopedia.com, Retrieved 5 January, 2019

- Cacciola, S.A.; et al. (2003). "First Report of Phytophthora palmivora on Grevillea spp. In Italy". Plant disease. 87 (8): 1006. Doi:10.1094/PDIS.2003.87.8.1006A

- Ralph C. Hassig; Kong Dan Oh (2009). The Hidden People of North Korea: Everyday Life in the Hermit Kingdom. Rowman & Littlefield. P. 298. ISBN 978-0-7425-6718-4. Retrieved 3 May 2015 – via Google Books

- Cultivar, definition: maximumyield.com, Retrieved 6 February, 2019

- "Vanda Miss Joaquim (Papilionanthe Miss Joaquim)". National Parks, Singapore Government. Retrieved 20 July 2014

5

Ornamental Plant Pathogens and Diseases

Various pathogens such as cymbidium mosaic virus, orchid fleck virus, diplocarpon rosae and podosphaera pannosa can cause diseases in different ornamental plants. The topics elaborated in this chapter will help in gaining a better perspective about these pathogens and diseases which can afflict various ornamental plants such as roses and orchids.

Diseases of ornamentals crops may be caused by pathogenic bacteria, fungi, viruses, andphytoplasmas. The occurrence of plant disease is a result of a complex interaction of a susceptible host-plant, the presence of a pathogenic causal organism, and the environment. The diseases result in hugeeconomic loss to farmers and some diseases are devastating in nature like wilt in carnation and gladiolus. Moreover, disease infected flowers are restricted to export and not accepted in domestic market.

Orchid Diseases

Leaf Spot [Gloeosporium sp., Colletotrichum sp., Cercospora sp. and Phyllostictina sp.]

The fungi produce dark brown spots of varying size on leaves. Warm humid weather and lack of light favours survival of the pathogen.

Removal and destruction of infected leaves prevent the disease from spreading. Spraying with Dithane M-45 (0.2%) or Bavistin (0.1%) is effective.

Pythium Black Rot [Pythium Ultimum]

The fungus affects the seedlings, especially under humid weather conditions. Affected plants turn black and leaves start falling.

Diseased leaves and plants should be removed and destroyed. Withholding of irrigation for few days and shifting the plant to less humid part help to check the disease. Fungicides such as Metalaxyl (0.1%), Fosetyl- Al or Mancozeb (0.2%) should be applied to control the disease.

Heart Rot [Phytophthora Palmivora]

The pathogen attacks the pseudo bulbs which results in yellowing and dropping off leaves. Pseudo bulb when cut open show dark rotted areas. This disease occurs on species of *Cottleya, Phalaenopsis* and Vanda.

Rogue out the infected plants. Spray systemic fungicides like Fosetyl- Al and Metalaxyl to prevent spread of the disease.

Brown Speck and Blight of Flowers [Botrytis Cinerea]

The genera Cattleya, Phalaenopsis, Dendrobium, Oncidium and Vanda are known to be susceptible to the disease. Infection appears on petals as small, water-soaked, brown spots that enlarge rapidly. In the advanced stage the entire flowers and some of the leaves may become covered with a grayish mold.

To keep the disease under control avoid excess humidity, keep the temperature high and provide good ventilation. Diseased plant parts should be removed and destroyed. Spray plants with Dithane M-45 (0.2%) or Bavistin (0.1%) at periodic intervals.

Rust [Hemileia Americana]

Rust caused by the fungus Hemileia Americana, is characterized by the appearance of orange yellow pustules on the underside of the leaves. Yellow green chlorotic areas also appear directly above the pustules on the upper surfaces.

Remove and burn all infected leaves. Dust the leaves with Sulphur or spray with Wettable Sulphur.

Orchid Wilt [Sclerotium Rolfsii]

The initial symptom is yellowing of leaf base which later turn brown. Infection of pseudo bulbs result is rotting and death of entire plant.

All infected leaves should be picked off and burned as soon as they are noticed. *Trichoderma* sp. has been found as very effective against the disease.

Bacterial Soft Rot [Erwinia Carotovora]

The symptoms appear as small, circular, blister like dark green colour spots on the upper surface of the leaves. The infection results in soft, pulpy and ill-smelling rot of the pseudo bulbs. It is a serious disease of *Cattleya* orchids.

Streptocycline (0.1%) is recommended for soft rot management.

Bacterial Brown Spot [Xanthomonas Cattleyae]

The disease occurs on plants of all ages. The initial symptoms appear on leaves as small watersoaked spots. The spots enlarge and run together and kill a large portion of the leaf, which then drops off. The water-conducting vessels may be invaded by the pathogen so that entire plant collapses.

All the infected plants should be removed and immediately destroyed. The spread of the disease can be prevented by repeated spraying with Agrimycin/ Tetracycline.

Viral Diseases

A number of virus diseases affects orchid. Among the most common are Cymbidium mosaic virus (CyMV), Tobacco mosaic virus-o (TMV-o) and Odontoglossum ring spot virus (ORSV).

These viruses cause a leaf necrosis expressed by rings, streaks, and irregular color patterns (mosaic). In severe cases malformation of leaves and flowers, mottling and twisting may result.

Remove all the diseased plants as they and spotted. Insects should be kept under control throughout the growing season. Spray with Malathion to control the several species of aphids and other insects that transmit the disease. Use virus free planting material for propagation.

Cymbidium Mosaic Virus

Cymbidium mosaic virus (CymMV) is a plant pathogenic virus of the family Alphaflexiviridae.

Cymbidium mosaic virus and the Odontoglossum ringspot virus (ORSV) are two of the most common viruses affecting cultivated orchids worldwide. Infected plants can have less desirable flowers or other problems, causing significant financial losses to orchid growers. The virus has not often been reported in wild orchid populations. It can be found in a wide variety of orchid genera but does not infect plants other than orchids.

Once an orchid is infected, the virus spreads throughout the infected plant in a number of weeks. Control measures may include sanitizing pruning equipment between plants. There is an ELISA test available to test for presence of the virus.

It is related to the Narcissus mosaic virus (NMV), the Scallion virus X (SVX), the Pepino mosaic virus (PepMV) and the Potato aucuba mosaic virus (PAMV).

Importance

Orchids are composed of nearly 200, 000 species of plants with attractive flowers. They are predominantly found in wet climates. The orchid family is one of the most important plant families in respect to the ornamental flower industry. In 2005, the potted orchid industry brought in about $144 million in the United States. This makes them the second most valuable potted plant in the nation. From 1996 to 2006 there has been a 206.4% increase in potted orchid prices. Globally, Taiwan, Taipei, Thailand, the United Kingdom, Italy, Japan, Brazil, and New Zealand are among the largest importers of potted orchids. Additionally, orchids are the source of vanilla. It is the only commercially important derivation from the plants - most commonly found in the Vanilla planifolia species. Certain orchids are also used in homeopathic treatments.

The virus stunts the growth of orchids both through size reduction as well as lowering flower yield. CymMV has also been linked to cases of breakage in flower coloration as well as blossom necrosis. This is especially important because this causes the commercial value of the plants to be greatly reduced.

Signs and Symptoms

CymMV causes a mosaic of irregularly shaped chlorotic and/or necrotic lesions to appear on infected hosts. Additionally, infected plants will show smaller yields. Sometimes the orchids may display chlorotic rings while others will display symptoms in lesions. Sunken patches can also be observed on leaves. Lastly, certain infected plants may be symptom-less but are still viral enough to infect other neighboring plants. This is because the chlorosis and necrosis can take time to show; however, the virus can still be present on a leaf that does not display such obvious symptoms of the disease. The virus is still able to be transmitted from the plant despite its "healthy" appearance. The virus can be detected with ELISA, immunodiffusion tests, or a nucleic acid hybridization assay.

Viral Cycle

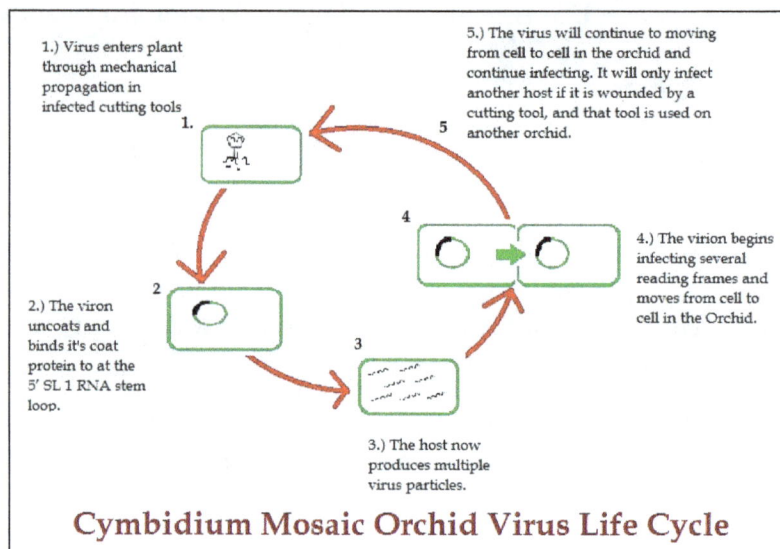

1.) Virus enters plant through mechanical propagation in infected cutting tools

2.) The viron uncoats and binds it's coat protein to the 5' SL 1 RNA stem loop.

3.) The host now produces multiple virus particles.

4.) The virion begins infecting several reading frames and moves from cell to cell in the Orchid.

5.) The virus will continue to moving from cell to cell in the orchid and continue infecting. It will only infect another host if it is wounded by a cutting tool, and that tool is used on another orchid.

Cymbidium Mosaic Orchid Virus Life Cycle

The Cymbidium mosaic virus starts out by infecting the host through a wound on the plant. Generally this wound occurs through plant propagation using contaminated cutting tools. Once inside the orchid, the virion uncoats, binds its coat protein to the host on the 5' SL 1 RNA stem loop and begins incorporating its RNA into it. This allows for the host cell to begin producing virus particles. The virus then begins infecting reading frames such as TGB1, TGB 2, and TGB 3, each of which aid in the virus's ability to spread from cell-to-cell in the orchid. When the virus infects TGB 1, it allows for the virus to move easily through the plasmodesmata. After the virus infects TGB 2 and TGB 3, it allows for the virus to begin moving through each of the orchid cells' endoplasmic reticulum. Affecting these sites allows for the virus to move easily through each of the plant cells. Overall, the monopartite, positive, sRNA from the virion connects its 3' tail with that of the host's 5' RNA. Translations will continue to occur, and disassemble from the 5' end of the virion. With more of the virions circulating throughout the host, it only takes another wound from a cutting tool to begin infecting another orchid.

Environment

When cultivated in a greenhouse or commercially, the virus is spread mechanically if tools used in cultivation are not properly sanitized. The virus exists systemically within the plant, and persists in sap that can be spread by such methods. In the wild, the virus is most likely spread by insect vectors. A different orchid virus, the orchid fleck virus, may be transmitted by the *Brevipalpus* mite. There is also evidence that cockroaches are able to transmit CymMV.

Management

There is no way to cure a plant that has the virus. The only thing to do once a plant is infected is to destroy it. The best management solution is to prevent the spread of disease. This is accomplished through effective disinfection of tools used in cultivation, including any plastic containers and razor blades. Autoclaving, flaming, and chemical treatment with trisodium phosphate and bleach solution are traditional methods for disinfection. The use of *Streptomyces* culture filtrate, which has also been shown to disinfect mad cow disease-causing prions, is another promising disinfectant. This method is shown to be effective in removing the virus from various tools, human nails, and orchid seeds. Another possible management strategy is development of plant resistance. An attempt to transform a mutant CymMV movement protein gene into *Dendrobium* orchids was slightly successful (9 of 259 plants were resistant and expressed the marker gene); the success of transformations actually conferring resistance appears to be related to a post-transcriptional gene silencing mechanism.

Orchid Fleck Virus

Orchid fleck virus (OFV) is a non-enveloped, segmented, single-stranded (ss) RNA negative-strand virus, transmitted by the false spider mite, *Brevipalpus californicus*. OFV causes necrotic and chlorotic lesions on the leaves of many genera in the family Orchidaceae.

Orchid fleck virus, despite its presence worldwide, only affects a small spectrum of human life. Orchids are not used for food but rather serve mainly as ornamental decoration. Therefore, only about 2 to 3 scientific reports are written about OFV each year. The whole genome of OFV has been sequenced and its six main protein products have been sequenced as well.

Much is still not known about OFV including how exactly and why vector mites travel from orchid to orchid, and more host species of flowers are being discovered annually. Many orchid enthusiasts are participating in "citizen science" by posting their OFV-related findings on international horticulture blogs and forums.

Virion Morphology

OFV was first described as bacilliform but depending on the method of preparation, OFV can appear bullet-shaped or bacilliform. On average, OFV is 40 nm in diameter and between 100 and 150 nm long. Each viral particles is organized into a tight 25 turn helix, with a pitch of about 4.5 nm.

Genomics

OFV contains two ssRNA molecules, RNA1 and RNA2, of 6413 and 6001 base pairs, respectively. GenBank contains the whole sequenced genome of OFV. RNA1 (GenBank AB244417) codes for five proteins whereas RNA2 (GenBank AB244418) only codes for one. Both strands possess open reading frames (ORF), which are read in the negative sense.

RNA1

- ORF1: Nucleocapsid (N) protein,

- ORF2: Phosphoprotein (P),

- ORF3: Proteins involved in viral cell-to-cell movement,

- ORF4: Matrix (M) protein,

- ORF5: Glycoprotein (G).

RNA2

- ORF6: Polymerase (L) protein (RNA-dependent RNA polymerase).

Pathophysiology

Chlorotic and necrotic flecks, spots, and/or ringspots, as well as yellow flecks or spots are all symptoms of an OFV infection. Studies have also shown that OFV may prevent the propagation of other viruses in an already OFV-infected plant.

Vector

The false spider mite, *B. californicus* serves as the major vector for OFV. *Brevipalpus* mites go through four distinct, active life stages, each separated by nonmotile chrysalis stages. The protonymph, deutonymph, and adult stages can infect their host plants with OFV, whereas the larval stage is not infectious. Even after three weeks of incubation of an OFV-positive mite on an OFV-resistant plant, *B. californicus* proved to still be infectious, showing that OFV is persistent.

Hosts

OFV is able to naturally infect around 50 different species in 31 genera, all belonging to the family Orchidaceae. The 25 other species from 11 non-orchid families have been infected through sap transmission or artificial viral inoculation.

Life Cycle

Studies have not shown whether or not OFV actually replicates within *B. californicus* but electron microscopy has revealed an intricate viral life cycle within the host cells.

Viral ssRNA is replicated and transcribed into mRNA in the host cell's nucleus. Viral mRNA is then exported out of the nucleus into the cytoplasm where it is translated into viral protein by the host's ribosomes. The viral proteins then reenter the nucleus where they aggregate into a viroplasm. There, the various viral structural proteins assemble with both strands of ssRNA to form complete OFV particles. These particles often cluster in between the inner and outer nuclear membranes, causing visible projections which often evaginate into cytoplasmic vesicles. Electron microscopy has revealed clusters of viral particles positioned perpendicular to the inner nuclear membrane, the endoplasmic reticulum, as well as the aforementioned cytoplasmic vesicles, forming distinctive "spoked wheel" structures.

Effects on Fitness

Infected orchids don't bloom as well as healthy ones, affecting efficacy of pollination and fertilization. Also, the orchids that do bloom look lifeless making them less attractive on the cut flower market.

Epidemiology

Cases of orchid fleck virus or OF-like viruses have been reported in Australia, Brazil, China, Columbia, Costa Rica, Denmark, Germany, Japan, Korea, South Africa, and the United States, i. e. every continent except for Antarctica.

Due to the fact that viruses depend on their host cell for replication, OFV cannot be cultured independently. However, two non-orchid indicator hosts (plants used in research that show characteristic symptoms of specific viral infections) C. quinoa and T. expansa are commonly used for viral inoculation and isolation.

There are no known pathogens of OFV itself but its vector, B. californicus has a symbiotic relationship with bacteria of the genus Cardinium. The symbiont is the cause of the mites' thelytokous method of reproduction (where females are produced from unfertilized eggs) and the explanation for the absence of male B. californicus mites.

Diagnosis

Thin tissue samples from plants with visible symptoms of OFV can undergo:

- Electron microscopy to visualize virions and complete viral particles.
- Serological analysis to identify and isolate specific viral proteins.
- Reverse Transcription-Polymerase Chain Reaction to identify viral RNA.

Prevention

Methods for preventing the spread of OFV among separate plants:

- Ensure seedlings are virus-free.
- Improve quarantine measures.
- Eliminate sources of infection (mites or other infected plants).
- Ensure proper environment for cultivation.
- Work towards developing OFV-resistant plants through genetic engineering.

Treatment

Once a plant is infected with OFV, it is unclear whether pruning visibly infected tissue will cure the plant of the virus. It is also unclear whether infected plants produce seeds containing viral particles.

Rose Diseases

Powdery Mildew [Sphaerotheca Pannosa var. Rosae]

The disease appears whenever the days are warm and nights are cool. The disease first starts on the young leaves as raised, blister like areas that soon become covered with a grayish, white, powdery fungus growth. The infected leaves are usually more purplish than healthy ones. New shoots get distorted. In severe infection the tips of canes may be dried. Frequently the unopened buds become white with mildew attack. Infected buds do not open and in open flowers the infection leads to discoloration and distortion of the petals. The overall plant vigor is checked. The primary mode of perennation of fungus is by infection of dormant buds. The pathogen produces conidia in chains which are readily dispersed by air and spread the disease.

The disease can be controlled by spraying Bavistin or Benlate (0.1%) at 30 days interval regularly. Wettable sulphur (0.2%), propiconazole (0.1%) and Karathane (0.05%) may be used at 7-10 days interval to control the disease. Wettable sulphur should not be used when day temperature exceeds 30 °C. The fungal antagonist Sporothrix flocculosa has bean reported to be highly effective in controlling the disease. Tilletiopsis pallescens, a naturally occurring ballistospore forming yeast isolated from mildew infected leaves has bean proved to be effective against the fungus. The bio agent reduces the incidence of disease by 97-98%.

Die Back [Diplodia rosarum]

Die back is one of the major diseases of rose. Disease appears in maximum severity following pruning of canes after monsoon. The disease causes death of the plant from tip downwards. The older plants are more prone to attack as compared to younger ones. Brown discoloration of the disease is conspicuous when affected stems are split open. The pathogen gets entry into the host tissue through the minute injuries caused by digger wasp.

The common practice is to cut away the affected plants and burn it. The secateur should be disinfected with spirit and cut ends immediately coated with chaubatia paste containing 4 parts of copper carbonate, 4 parts of red lead and 5 parts of linseed oil.

Alternaria Leaf Spot [Alternaria alternate]

The disease causes heavy losses during the rainy season. Young leaves are more susceptible. Small oval to irregular, dull brown to black scattered spots are first noticed on the margin of leaves which later enlarge, become confluent and cover the entire leaf surface. Pathogen survives on diseased leaves and other plant parts. A temperature of 28-300C and high humidity is suitable for the attack of the disease.

Diseased leaves should be collected and burnt. Four sprays with Benlate (0.06%), Captan (0.25%) or Mancozeb (0.25%) at 10 days interval during December-January can effectively check the disease.

Rust [Phragmidium spp.]

The disease is characterized by the presence of reddish-orange pustules on leaflets and sometimes on petioles. The colour of these pustules changes to black when teleutospores are formed. In severe cases of infection, defoliation occurs. The production of flowers from diseased plants is drastically reduced. The pathogen overwinters in the form of teliospores or as mycelium in the infected stem. Mild winter temperature and rainfall are most suitable conditions for rust epidemics.

Fallen affected leaves should be collected and destroyed. Spring pruning and dormant spray of Copper oxy chloride (0.3%) is effective in controlling the disease. The disease can be effectively controlled by spraying Dithane M-45 (0.2%), Vita vax (0.1%) or Benodonil (0.1%) three times at 15 days interval during March-April. Dithiocarbamates, oxycarboxin and ergosterol biosynthesis inhibiting fungicides have provided good control of rose rust.

Botrytis Bud and Twig Blight [Botrytis Cinerea]

The fungus mainly attacks flowers and flowering stems. The symptoms are noticed as brownish patches on petals of flower buds which enlarge and cover the entire surface. The infected buds droop down. The sunken grayish black lesions extend to the stem from the base of the bud. Die-back like symptoms are visible sometimes. The fungus overwinters in the form of mycelium or sclerotia in the infected plant debris. Conidia are airborne and cause the spread of the disease. Prolonged periods of free moisture favour infection.

Spraying with Bavistin (0.1%) or Mancozeb (0.25%) has been found to be useful in controlling the disease. Sanitation practices have the potential to reduce the incidence of disease. All fallen leaves, bloom and other plant debris should be removed and destroyed. Trichoderma viride (PDBCTV 4) is effective in controlling the disease. Garlic juice at 0.5-5% conc. decreases development of botrytis blight.

Anthracnose [Sphaceloma Rosarum]

Small, circular, brown spots with purplish border appear on leaves which increase in size and cover the entire lamina and severe infection leads to complete defoliation. In some cases the diseased tissue falls out, giving the leaves a shot-hole appearance.

Collect all affected leaves and destroy by burning. The disease can be successfully controlled by spraying of Benlate (0.1%).

Rose Mosaic Virus (RMV)

Rose mosaic occurs mostly on green house roses. Disease is characterized by yellow or whitish chlorotic lines, rings, mottles or net like mosaic patterns on the foliage. The infected plants develop general decline, dieback and premature defoliation.

Diseased plants should be destroyed and never used for propagation. Disease free budding and grafting material should be used.

Diplocarpon Rosae

Diplocarpon rosae is a fungus that creates the rose black spot disease. Because it was observed by people of various countries around the same time (around 1830), the nomenclature for the fungus varied with about 25 different names. The asexual stage is now known to be Marssonina rosae, while the sexual and most common stage is known as Diplocarpon rosae.

Diplocarpon rosae over seasons as mycelia, ascospores, and conidia in infected leaves and canes. In the spring during moist, humid conditions, ascospores and conidia are wind-borne and rain-splashed to newly emerging leaf tissue. Upon infection, disease progresses from the lowest leaves upward, causing defoliation and black spots on leaves.

Diagnosis

The black spots are circular with a perforated edge, and reach a diameter of 14 mm. Badly affected plants, however, will not show the circular patterning, as they combine to cause a large, black mass. The common treatment of the disease is to remove the affected leaves and spray with anti-fungal solutions. Some stems of the roses may become affected if untreated, and will cause progressive weakening of the rose.

Disease Cycle

Diplocarpon rosae tends to overwinter in both lesions of infected canes and fallen foliage. Conidia are produced in the diseased stem tissues and dispersed via water—most commonly by rain or wind—into the openings of leaves in the spring season. The conidia then produce germ tubes (and sometimes appressoria) to penetrate the tissues of the leaves. Mycelia develop on the underside of the leaf cuticle and lesions appear. As these lesions appear, acervuli continuously produce conidia asexually as long as the climate remains optimally wet and warm. These conidia can then be dispersed to new uninfected leaves as a source of secondary inoculum, adding more cycles of infection. Once defoliation occurs in the fall season, the hyphae of the Diplocarpon rosae invade the dead leaf tissue and form pycnidia lined with conidiophores under the old acervuli. The pycnidia then overwinter in the lesions of infected tissue and burst in the spring, releasing conidia to be dispersed by water and effectively completing the disease cycle. Diplocarpon rosae also has a sexual stage, although this is rarely observed in North America due to unfavorable environmental conditions. In this stage, the sexual spores (ascospores) are formed in the apothecium. If the weather conditions are favorable for the formation of ascocarps, the apothecia that contain asci can be observed in the spring. However, this rarely occurs, and the fruiting bodies are typically filled with conidia that enable the asexual life cycle of the pathogen to occur.

Environment

Diplocarpon rosae typically favor environments with a warm and wet climate. Conidiospores involved with infection are only dispersed via water, making the disease most active in the late spring and early fall seasons, or other periods that experience similar climate conditions. The development of the Black Spot disease itself is ideal at temperatures ranging from 68–80 °F (20–26.7 °C). It is important to note that no infection will develop if the leaf surfaces dry out within 7 hours of the initial conidial germination. Similarly, temperatures above 85 °F (29.4 °C) also halt the spread of disease.

Treatment

Removing infected leaves from the plant and fallen leaves from the ground will slow the spread of the infection, as does avoiding wetting the leaves of plants during watering. An infected plant can be removed from the area, which will slow the spread of infection to other plants, but this often is not desirable. Fungicides, such as mancozeb, chlorothalonil, flutriafol, penconazole, or a copper-based product, applied upon new leaf emergence or first appearance of black spot, can be used to control the disease. If a more natural and nontoxic approach is desired, diluted neem oil is effective both against black spot and as an insecticide against aphids. It is usually necessary to repeat the spraying at seven- to 10-day intervals throughout the warmest part of the growing season, as the fungus is most active at temperatures from 24 to 32 °C (75 to 90 °F).

Importance

Black Spot of rose is the single most impactful disease of roses globally. Every year around 8 billion flowering stems, 80 million potted plants and 220 million garden rose plants are sold commercially. All species of roses (Hulthemia, Hesperrhodos, Platyrhodon and Rosa) are affected by black spot disease. The disease is found everywhere roses are planted, typically in epidemic proportions. The water-borne dispersal methods allow it to infect a plethora of plants every growing season and increase the overall incidence of disease. Although Diplocarpon rosae does not kill the rose itself, it is known to completely defoliate the leaves of the rose plant. This is a huge issue when dealing with such an aesthetically commercialized crop such as the rose. Additionally, the weakened rose plant will become more susceptible to other pathogens and disease following infection.

Podosphaera Pannosa

Podosphaera pannosa is a plant pathogen. It produces a powdery mildew on members of the rose family.

Rose powdery mildew [also known as 'Weeping Mildred'] is caused by the fungus *Podosphaera pannosa*, a member of the Ascomycete fungi. It infects a wide variety of roses, but especially those grown in dryer climates as the fungus has the rare characteristic that not only does it not need water to germinate and reproduce, it can be inhibited by it.

Disease Cycle

The disease cycle of rose powdery mildew starts when the sexual spores, ascospores, of the pathogen survive the winter in a structure composed of hyphae called an ascocarp. The specific ascocarp is a chasmothecium, or cleistothecium, and has a circular shape to it. Under the right conditions the chasmothecium will break open to reveal the asci, which are long tube-like structures containing the ascospores. These ascospores are then released and spread by wind, insects, and rain until they land on a susceptible rose for a host and land, attach, and germinate on the plant. They will also produce condia, the asexual spores of *Podosphaera pannosa*, which will spread throughout the summer. It is these long chains of white conidia which give the fungus its characteristic "powdery" appearance. Late in the year as the plant is dying cleistothecia will again form when the ascogonium receives the nucleus from the antheridium.

Environment

Optimal conditions for rose powdery mildew are 16-27 °C, with the optimal temperature for fungal growth at 23 °C in a shaded area. They also do not need water to germinate and infect the rose. In fact, if there is too much water present on plant surfaces for a prolonged period of time the fungal growth is inhibited and the spores can actually die. Rose powdery mildew can also grow in any conditions where roses can grow and has been found everywhere from China to the United States.

Hosts, Signs and Symptoms

A wide variety of rose species are susceptible to powdery mildew. In light of this it is more practical to discuss the rose varieties that are resistant as opposed to those that are susceptible. Two varieties that have been found to show resistance to rose powdery mildew are "Paul's Pink" rose variety and the "Magic" rose variety. Other research has shown that many chestnut rose (Rosa roxburghii) varieties are also resistant to powdery mildew. Rosa sterilis, Rosa kweichowensis, Rosa laevigata, Rosa lucidissima, and Rosa chinensis have all been shown to be resistant to powdery mildew. R. multiflora var. multiflora and R. multiflora var. cathayensis have all been shown to be susceptible to rose powdery mildew. Symptoms caused by the rose powdery mildew can be a dwarfing of the growth of the plant, or the twisting and deforming of leaves, but more noticeable is a sign of the disease, which is the white condia, the "powder" that appears on the plant surfaces, such as leaves, shoots, flowers, and buds. The fungus may grow on both new and old leaves, but is generally more concentrated on the underside of the leaf.

Management

Effective management of rose powdery mildew begins by using resistant varieties of rose, but it can also be managed through the use of fungicides, or by planting in sun since rose powdery mildew prefers the shade. In fact, increasing the exposure of rose powdery mildew from 18 to 24 hours of light per day reduced the production of conidia, the asexual spores of the fungus, by as much as 62%. There are a variety of fungicides that have proven to be effective. Examples are myclobutanil, azoxystrobin, triadimefon, and thiophanate-methyl formulations Chemical fungicides are not always necessary, however, it is possible to use more environmentally-friendly solutions such as a water-vinegar spray, or mixtures of baking soda and insecticidal soaps. Recent studies have also shown that using a planting medium which includes silicon can also reduce the occurrence of powdery mildew by as much as 57%.

Importance

Powdery mildew affects more the 7600 species of hosts worldwide, including subsistence crops. Although rose powdery mildew will most directly affect the rose connoisseur, it is part of this larger family of powdery mildews, which can affect the crops used for food and survival in many countries, thereby having economic and human impacts beyond that of an unsightly rose bush. Research shows that total yield loss from powdery mildew on cereal crops alone can vary anywhere between 2 and 30% depending on the host and the environmental conditions. Additionally, the wholesale value of roses annually exceeds $100 million in the United States, so the national economic impact from the flower industry cannot be ignored.

Other Plant Diseases

Diseases of Gladiolus

Fusarium Rots and Yellows [Fusarium Oxysporum F. sp. Gladioli]

It causes curving, bending, arching, stunting, yellowing or drying of leaves associated with root and corm rot in field as well as in storage. The disease is characterized by dry rot in storage, often restricted to the corm base. Corms when split into halves, show radiating dark- colored streaks extending from the corm base through the flesh and in severe cases the centre of entire corm is black and rotten. Initial infection of the corms comes either through the soil or by latent corm infection of previous year which in adverse weather conditions develops bright yellow interveinal streaks on leaves. A plant bending geotropically with leaf tip burn is a conspicuous feature. The bending always starts from the site of the corms showing rot. The initial infection of the corms comes either through soil or by latent corm infection of the previous year. Fungus is carried practically in all stocks of corms and cormels as latent infection. The pathogen is also disseminated through soil, contaminated water and leaf hoppers.

Water soaking of cormels at 57.2 °C up to 30 minutes reduces the disease provided they had been lifted in mid April, or early summer and treated 6 to 10 weeks after harvest. Pre planting or post harvest dusting of corms with 10 to 20 per cent Benomyl or 10 to 20 percent TBZ (Thiabendazole) is very effective. Disease can be effectively controlled by dipping of corms in 0.1 per cent Carbendazim before planting. Plant extract of Allium sativum and Ocimum sanctum has provided 61-67 % disease control.

Core or Spongy Rot [Botrytis Gladiolorum]

Disease starts as small spots which develop into large spots covered with mould, soon killing the leaves. Characteristic symptoms include small red-bordered and rusty colored specks on leaves and stems and pale brown to dark brown spots on the flowers. Through the vascular bundles of stem and leaves the fungus reaches the corms and causes core rot. It is a soil borne highly destructive fungus which affects the plants during cool and humid weather, most favorable temperature being 13 to 15 °C.

The disease can be controlled effectively by spraying fungicides such as Vinclozolin and Benlate. Mancozeb sprays @ 0.2% also control its infection. Hot water treatment (52 ° C) of corms is effective in eradicating the pathogen. Corms should be dried well before storage.

Dry or Neck Rot [Stormatinia Gladioli]

The fungus produces sclerotia frequently but apothecial formation is not common in nature. The infection originates from infected corms and sclerotia. High proportion of buried sclerotia in soil increase spread of disease by close planting. Dry rot is a wide spread disease attacking gladioli plants in the field and corms in storage. It is more severe during humid conditions in the field where premature yellowing occurs ultimately causing death of the plants. The leaves turn brown from the tips downward and at bases they decay causing neck rot but corms remain attached to the stem firmly. Diseased corms show numerous round black and small lesions varying in size with slightly raised edges which may coalesce together forming irregular area.

Four to six years crop rotation is essential to remove the soil borne inoculums. Hot water treatment is effective in eliminating the pathogen. The disease can be managed by soaking the cormels first in cool water for 24 hours and then for 30 minutes in hot water at 54.5 ºC. Corms should be treated with Thiram or Dichloran (0.3%).

Storage Rot [Penicillium Gladioli]

The disease mostly appears in the form of black, brown, greenish or yellowish mouldy growth on the corms during storage. Under poor air circulation the corms may rot and emit a foul smell.

As infection occurs through injuries, wounding of corms at the time of digging or during handling should be avoided. The corm after digging out from the soil should be treated with Thiram 75 DS (0.3%) and properly cured at 30-35 °C for 7-10 days before they are taken to the cold store. Damp storage conditions should be avoided. High temperature (more than 5 °C in the cold store) also needs to be avoided, as it could lead to rapid rotting of corms under humid conditions.

Scab [Burkholderia Gladioli Pv. Gladioli]

On the leaves, the symptoms appear as brownish yellow specks. On the corms, the symptoms appear usually as circular, brown, sunken lesions with raised margins. They develop more abundantly on the lower half of the corm. The neck rot type of symptoms appear as numerous, brown to black, small spots near the base of the plants. The lesions often ooze out gum like exudations. The pathogen multiplies fast and spread through soil or may be carried over with the corms. It can be transmitted by mites and root knot nematodes.

Since the disease can be transmitted by mites, nematodes and other soil insects, insecticides such as Thimet, Furadan or Temik may be applied in open furrows at planting before covering the corms with soil. Corms and cormels should be soaked in Formeldehyde (0.5%) for 2 hours then in 200 ppm Streptomycin for 2 hrs just before planting. Dipping of corms in 0.2% Thiram suspension for 15 minutes also checks infection.

Curvularia Leaf Spot [Curvularia Trifolii F. sp. Gladioli]

The pathogen is found in the soil. Disease spreads rapidly under moist conditions. This disease mostly affects leaves but other plant parts are also invaded. On leaves, the spots are initially small and round, but later extend mainly in the direction of veins. They are light brown, surrounded by a darker reddish brown ring and a yellowish halo. Black spore masses of the fungus form in a scattered fashion in the centre of these spots. On stem, the spots are brown, elongated and sunken. These may girdle the stem and result in breaking of the plant above the point of infection. Brown to black lesions appears on the corms also.

Healthy disease free corms should be used for planting. The disease may be controlled by spraying Dithane M-45 @ (0.2%).

Root Knot [Meloidogyne Incognita and M. Hapla]

The affected plants show retarded growth and turn pale in colour. The characteristic symptom is formation of galls on roots of the plant.

Deep ploughing in summer and crop rotation of at least 3 years should be practiced. Thimet 10 G or Temik 10 at the rate of 12 kg of granules per acre should be worked into the soil.

Viral Diseases

Several viruses are known to infect gladiolus. Most of them are transmitted by aphids or nematodes. Aphids transmit mostly mosaic type viruses (for example, cucumber mosaic and bean yellow mosaic virus), whereas nematodes the ring spot viruses (for example, tobacco ringspot and tomato ringspot virus). Aster yellow is reported to be transmitted by leaf hoppers. Flowers are small, distorted or have colour breaking symptoms. Leaves may be mottled, have white flecks or reddish blotches.

Use of virus free planting material is highly beneficial to raise healthy plants. Meristem tip culture is known to eliminate most viruses form infected plants and hence of value where whole of the planting stock is infected. Healthy planting stock must be maintained under insect-proof structures or by controlling insect vectors with suitable insecticides. Genetic resistance of the host to the major virus is a long term approach to the problem. Soil solarization should be done to remove nematodes and weeds. Infected plants should be destroyed as soon as they are noticed.

Diseases of Carnation

Wilt [Fusarium Oxysporum F. sp. Dianthi]

The initial symptoms of the disease are yellowing of foliage and production of crook neck shoots. The stems are softened so as to be easily crushed. Stem when cut open show brown zonation or striping at vascular region. The entire plant wilts and collapses in a very short time.

Since the fungus involved in this disease is soil born, the first point in control is to avoid contaminated soil. Drenching of soil with Copper oxychloride (0.4%) and spraying with Bavistin (0.1%) reduces the malady. Soil solarization has been found very effective in minimizing the disease. Biological agents like Trichoderma harzianum, Pseudomonas fluorescens, Bacillus subtilis, Streptomyces sp. and non-pathogenic isolates of Fusarium are reported to be effective against the disease. Neem based formulations have also been reported to be effective against the disease.

Foot Rot [Phytophthora Nicotianae var. Parasitica, Pythium sp. Rhizoctonia solani, Sclerotinia sclerotiorum]

Under high moisture condition fungi attack the root and collar portion of the stem at soil level which later results in wilting. The leaves get discolored and start drying up from bottom upwards.

Avoiding excess soil moisture is an important step in the management strategy. Soil treatment with Benlate or Iprodione against Rhizoctonia solani and Fosetyl-AI and Metalaxyl against Pythium and Phytophthora are effective. Biological control by Trichoderma harzianum has been found to reduce the disease incidence by 70%.

Leaf Spot [Alternaria Dianthi]

The disease produces light brown lesions with purplish brown border on leaves and stem. The lowest leaves are attacked first and the disease progresses upwards. On enlargement, the lesions merge and result in blighting and premature death of leaves.

Remove infected leaves. Foliar application of Dithane M-45 (0.2%) or Bavistin (0.1%) is effective to minimizing disease losses.

Fairy Ring Spot [Mycosphaerella Dianthi]

The fungus causes fairy ring like spots which coalesce, extend and merge, eventually destroying the leaves.

Spraying of Dithane M-45 (0.2%) or Bavistin (0.1%) at an interval of 10 days controls the disease.

Basal Rot [Sclerotium Rolfsii]

The pathogen infects the stem at the soil level and which later extends to leaves causes stem rot.

Cottony growth of the fungus mycelium is seen on the infected portion.

Remove old and senescent leaves touching the soil. Regulation of soil moisture and avoiding excessive nitrogen to the soil helps to control the disease. Drench the soil near the stem base with Thiram (0.25%) or Carbendazim (0.1%).

Rust [Uromyces Dianthi]

Long narrow chocolate-brown colour pustules can be seen on leaves, stems and flower buds. Infected plants become stunted and their leaves curl up.

Keep the foliage dry. Spray Dithane M-45 (0.2%) or Bavistin (0.1%). *Verticillium lecanii* significantly reduces the incidence of disease.

Phialophora Wilt [Phialophora Cinerescens]

The pathogen results in gradual wilting of the plant. Foliage of the infected plants fade and turn into straw color. Stems when cut open show chocolate brown discoloration at the vascular region. In advanced stage of infection wilting of whole plant can be noticed. Control measures are same as in Fusarium wilt.

Gray Mold [Botrytis Cinerea]

The disease is very common during stage of cut-carnation. Under high humidity conditions the fungus causes water soaked flecking of the outer petals and gradually the entire flower is affected.

Soil drenching and spraying of plants with Bavistin (0.1%) minimize disease intensity. Mycostop biofungicide (*Streptomyces griseoviridis*) has been found effective in minimizing the disease.

Stem Rot [Fusarium Roseum. F. sp. Cerealis]

The disease occurs due to injuries caused by continuous harvesting. Rotting of the stem at the soil level or higher on the plants is seen quite often. Dry rot at the base causes the plant to wilt.

Since the disease is soil borne, soil fumigation and soil solarization have been found effective in controlling the disease. Spraying of Benomyl or Bavistin (0.1%) also controls the disease.

Bacterial Wilt [Burkholderia Caryophylli]

Foliage becomes grayish green then yellow and finally wilting of the plant may occur. Vascular discoloration, bacterial ooze and rotting of roots are the other symptoms. Plants can be easily pulled out of the soil.

Disease free planting material should be used. Soil should be sterilized. Rouging and removed of symptomatic plants. Overhead irrigations should be avoided.

Diseases of Marigold

Damping Off [Rhizoctonia Solani]

Symptoms appear as brown necrotic spots, girdling the radical which later on extend to plumule and cause pre-emergence mortality. Post- emergence symptoms appear on lower part of hypocotyls as water soaked, brown, necrotic ring leading to collapse of seedling when infected seedling are pulled, the root system appears partially or fully decayed.

Proper drainage should be provided in nursery bed. Soil drenching with Carbendazim (0.1%) should be followed to manage the malady. Three-four years crops rotation should be followed.

Leaf Spots and Blight [Alternaria spp. (A. Tagetica, A. Alternata), Cercospora spp. Septeria spp.]

Minute Brown, circular spots appear on lower leaves which enlarge at later stage of infection leading to premature defoliation, reduced flower size and ultimately death of the plants.

To keep the disease under check, the marigold crop should be sprayed with Dithane M-45 @ 0.2% or Carbendazim (.05%) at fortnightly intervals starting from the first appearance of disease symptoms.

Powdery Mildew [Oidium sp., Leveillula Taurica]

Whitish, tiny, superficial spots appear on leaves which later on result in the coverage of whole aerial parts of plant with whitish powder.

The disease can be controlled by spraying with Karathane @ 0.05% or Sulfex (3g/l of water) at fortnightly intervals.

Flower Bud Rot [Alternaria Dianthi]

Disease appears mainly on young flower buds and results into their dry rotting with brown scorched, necrotic discoloration of sepals and stalk. The ray and disc florets also turn brown. At later stages buds become shriveled, turn dark brown and dry up. Symptoms are less prominent on mature buds but these buds fail to open. Brown necrotic spots are visible on margins and tips of older leaves.

To control this disease regular spraying of Dithane M-45 @0.2% should be followed.

Diseases of China Aster

Wilt [Fusarium Oxysporum F. sp. Callistephi, Verticillium Albo-atrum]

Disease is characterized by stunted growth, yellowing of leaves followed by withering of plant and subsequent rotting of the collar region. When the stem of infected plant is cut, the vascular ring is found to be brown, especially on the most affected side. Such plants finally wither.

Since the fungus over winters in soil and may be carried on all parts of the diseased plants, the seeds collected from diseased plants must be disinfected as well as the soil in which they are to be planted. Soaking of seeds for 30 minutes in a 0.1% solution of Mercuric chloride and steam sterilization of soil has proved very effective to prevent the disease.

Collar or Root Rot [Phytophthora Cryptogea]

Stems and roots of infected plant appear water soaked and black. The rot caused by the fungus is more pronounced and lacks the pink coating of spores characteristic of plants infected by wilt.

The control measures include strict limitation of irrigation and avoiding planting in the vicinity of alternative hosts. If possible, planting should not be done in the field which has shown the disease in previous years. If it is necessary to use the same beds, the soil should be sterilized.

Grey Mould [Botrytis Cinerea]

Symptoms appear in the form of blossom blight, bud rot, stem canker, stem and crown rot, leaf blight and damping off. In moist weather diseased tissue start rotting. On these tissues gray or brown coloured growth of the fungus can be seen. Disease spreads rapidly during cool and humid weather.

The disease can be reduced by using light well drained soil. Infected plants or plant parts and weeds should be removed and destroyed. Overcrowding of plants and overhead irrigation should be avoided. Seeds should be dried at 18-20 °C. Three-four spraying with Mancozeb (0.25%) or Chlorothalonil (0.2%) should be done.

Rust [Coleosporium Asterum]

Bright yellowish orange spots appear on the lower surface of the leaves, particularly on those of young plants. Initially these spots are covered with thin layer of host but on maturity they become erumpent exposing orange red colour powdery spore masses. Growth of diseased plants is reduced and quality of flower s is affected.

Infected plant parts should be collected and destroyed. Sprinkler irrigation of plants should be avoided.

Spraying of wettable sulphur during the growing season is effective to control the disease.

Leaf Spot [Ascochyta Asteris, Septoria Callistephi, Stemphylium Callistephi]

The spots are first yellowish, and then become dark brown and black, increasing in size. The lower leaves are infected first. In case of severe infection diseased blighted leaves fall on the ground. Quality of the flowers is reduced.

Diseased plant parts should be collected and burnt. Seed should be treated with Thiram (0.2%) or Carbendazim (0.1%) before sowing. Disease can be effectively controlled by spraying with Dithane M-45 (0.2%) at weekly interval.

Diseases of Chrysanthemum

Bacterial Blight [Erwinia Chrysanthemi]

The pathogen causes wilting of plants on bright days. With the spread of the disease, the stem tips turn brown, brittle and collapse. Stem becomes hollow with brownish streaks extending up to base.

The methods to minimize the disease include destruction of affected plants, soil sterilization, using disease free cutting and avoiding contamination during pinching. Spray of Streptocycline (0.01%) has been found to be effective.

Bacterial Leaf Spot [Pseudomonas Cichorii]

Infected plant leaf show dark brown to black slightly sunken spots having concentric zonation. On expansion, the spots merge and form large necrotic areas. In advanced stage the bacterium invade the flower buds which turn dark and die prematurely.

Use of disease free cuttings and spraying with Streptocycline (0.01%) is effective.

Root Rot [Pythium spp. and Phytophthora spp.]

Roots get destroyed. Diseased plants are stunted and appear pale yellow. The infected plants can be pulled out easily.

Among cultural practices soil solarization, removal of infected plants and good drainage facility are important. Fungicides recommended against the disease are: Metalaxyl, Mancozeb, Captan, and Fosetyl-Al.

Phoma Root Rot [Phoma Chrysanthemi]

The infected plants show stunting with a yellowing of lower leaves and some cracking of the main stem.

Drenching of soil with Copper oxychloride (0.3%) two weeks before planting offers good control. Soil treatment with *Trichoderma* isolates (T 8 B, T5B, T73B, T P5, T 40 and T 20B) has been found effective in controlling the disease.

Foot Rot [Rhizoctonia Solani]

The disease is very common in warm moist conditions. The attack of fungus results in mushy brown rot of stem and leaves.

Soil sterilization and drenching with copper fungicides give effective control of the disease. Bio-fungicide made from the combination of Bacillus subtilis and Kaslin has been reported to give protection against foot rot. It has been also reported that spraying with or dipping in Benomyl (0.2%) gives protection to young plants.

Stem Rot [Fusarium Solani]

The infected plants show necrosis of leaf and decay and discoloration of pith and the adjacent vascular region of the cortex. The fungus produces small dark lesions at the base of stem. In the advanced stages of infection root decay is observed.

Soil treatment with Dithane M-45 or Copper fungicide before planting is recommended to control the disease. Spraying with Bavistin (0.1%) is also effective to minimize the disease.

Wilt [Fusarium Oxysporum F. sp. Chrysanthemi]

The infection results in chlorosis and necrosis which start from lower leaves. Stem near soil level becomes black, and brown discolouration extends in the wood of the stem for a considerable distance above the ground. Finally the whole plant wilts and dies.

Soil treatment with Thiophanate methyl has been found effective in minimizing the disease.

Powdery Mildew [Erysiphe Cichoracearum]

Appearance of White powdery growth on the upper surfaces of the leaves is chief diagnostic symptom of the disease. The pathogen also attacks the stem and flower.

Spraying with Karathane (0.025%), Bavistin (0.1%) or sulphur based fungicide (0.2%) and providing a dry environment helps to check this disease.

Rust [Puccinia Chrysanthemi]

The disease is characterized by the appearance of small pinhead size blisters on the underside of the leaves. Blisters also appear to some extent on the upper surface. These blisters break open and expose a dark-brown, powdery mass of spores.

Remove diseased leaves from the plant as soon as noticed and burn them. Overhead irrigation should be avoided. Spraying with Mancozeb (0.2%) is effective in minimizing the disease.

Verticillium Wilt [Verticillium Dahliae and V. Albo- atrum]

The initial symptoms are yellowing and browning of lower leaves. Eventually, more leaves become affected and turn brown and die. Infected plants are stunted and often fail to produce flowers.

Regular heat or chemical treatment of the soil helps to minimize the incidence of the disease. Soil solarization can effectively minimize the disease.

Grey Mould [Botrytis Cinerea]

The pathogen attacks flower in moist green houses causing the development of brown water soaked spots. Infected parts become covered with a grayish brown powdery mass of spores.

A control measure consists of providing better ventilation and good aeration by providing adequate planting distance. Spraying with Bavistin (0.1%) and Copper oxychloride (0.2%) gives good results.

Leaf Spot [Septoria Chrysanthemi]

The symptoms appear on leaf as yellowish spots which later become dark brown and black. The spots grow in size and number, coalesce with one another and form large patches, covering a major portion of the leaf in humid weather. Serious infection may result in premature withering of leaves.

A control measure includes picking and destroying the infected leaves. The application of Bavistin (0.01%) or Benlate (0.01%) is effective in checking the infection. Excessive irrigation should be avoided.

Ascochyta Blight (Black Rot) [Ascochyta Chrysanthemi]

The infection results in development of blackish lesion on stem and the lower leaves which later grow bigger to make irregular black blotches. It also causes browning of petals and flower stalk which drop after becoming black.

Amongst the cultural method field sanitation is best to reduce the intensity of disease. Dipping the cutting in Benlate (0.1%) and spraying with Bavistin, Benlate or copper fungicide have been found to give good control of the disease.

Ray Speck [Stemphylium Lycopersici]

The infection results in brown or white necrotic specks surrounded by coloured halos on ray florets.

The attack is severe when humidity and temperature are high.

Application of Dithane M-45 (0.2%), Bavistin (0.1%) or copper oxychloride is done to control the disease. Avoid excess moisture by providing better ventilation in the green house.

Diseases of Lotus

Leaf Spot [Cercospora sp. and Ovularia sp.]

They cause dark patches on the pads and may eventually, if unchecked cause pad to die. Rarely is the entire plant affected. Implications for rhizome development would be through loss of photosynthates and thus lower yields.

Copper fungicides such as Bordeaux mixture applied to leaves and water helps to control the disease.

Root Rot [Phytophthora sp.]

The fungus causes crown and stem base to blacken and rot and it quickly spreads through the pond, with early symptoms indicated by yellowing of leaves. Further examination reveals black jelly like tissue and a foul smell even if roots appear healthy.

Infected plants should be removed and burnt. Some degree of protection can be achieved by impregnating water with Copper sulphate crystals dragged through water in an infusing bag. However, if plants do not respond, all plants should be destroyed and pond sterilized with Sodium hypochlorite.

Diseases of Jasminum

Leaf Blight [Cercospora Jasminicola, Alternaria Jasmini]

Disease spreads rapidly in the rainy season. Reddish brown circular spots, 2 to 8 mm in diameter are produced on the upper surface of the leaves. Affected leaf margins show inward curling and become hard and brittle. In severe cases of infection, vegetative buds and young branches dry up.

Spraying with Bavistin (0.1%) or Copper oxychloride (0.3%) at monthly interval commencing from May onwards up to pruning helps to control the disease. Diseased leaves should be collected and burnt.

Rust [Uromyces Hobsoni]

The leaves show the presence of orange coloured aerial cups on both sides, but predominantly on the lower surface. Numerous blisters are produced in advanced stage of infection causing yellowing and crinkling of the leaves. The stems and branches are also infected, causing splitting of barks and subsequent death of the branches.

Affected plant and plant parts should be removed. The disease can be controlled by dusting Sulphur @ 20-25 kg/hectare. Spraying with Bordeaux mixture or Copper oxychloride (0.3%) is also recommended for the control of the disease.

Wilt [Fusarium Solani, Sclerotium Rolfsii]

The early symptom is yellowing of lower leaves which gradually spread upwards and finally resulting in death of the plant. The disease occurs in patches. The infected plant base shows a network of fan-shaped mycelial strands of the fungus which later on produce mustard like brown sclerotia. In the case of sclerotial wilt, in addition to the above symptoms, white mycelia are found generally girdling the roots and the sclerotia are found adhering to the roots of the wilted plants.

Drenching the soil around the plants with 1% Bordeaux mixture prevents the disease spread. P. fluorescens, B. subtilis and T. viride have been found effective against sclerotial wilt. Soil application of talc based commercial formulation of P. fluorescens @ 20g/pot, B. subtilis and T viride @ 25g/pot effectively reduced the wilt disease incidence in the pot culture experiment.

Other Leaf Spots [Fulvia Fulva]

Symptoms are noticed on the upper portion of leaves in the form of light yellow spots which later turn to olive brown and infected leaves become blighted because of convergence of many spots.

Alternaria Leaf Spot [Alternaria Alternata]

Irregular brown lesions surrounded by dark coloured bands appear on leaf lamina which some-time result in "Shot hole Symptom".

Phoma Leaf Spot [Phoma Harbarum]

The infection appears as small circular light brown spots which later on coalesce and develop in to brownish to ashy necrotic areas. Shot hole symptoms are observed in old lesions.

Glomerella Leaf Blight [Glomerella Cingulata]

Irregular water soaked dark brown spots appear on leaves and twigs. Coalescence of spots on leaves and twigs results in blighting symptoms.

Management: Similar to leaf blight.

Mosaic [Virus]

The diseased plants show stunting and yellowish green appearance with small leaves. Yellowish green to chlorotic flecks of 1-2 mm in diameter appear irregularly on the leaf and these streaks form into a ring.

Control of insect vector with Metasystox (0.1%) prevents the transmission of the disease.

Phyllody [Phytoplasma]

The affected plants produce malformed, reduced greenish flower-like structure instead of fragrant white flowers on panicles which are highly congested and green in colour. The greenish corolla lobes become reduced and ovate in shape. Flower parts are transformed into leaf like structure.

The disease may be controlled by spraying with Tetracycline hydrochloride (250ppm). Cuttings from infected plants should never be used for planting.

Diseases of Tuberose

Foot and Tuber Rot [Sclerotium Rolfsii]

In the field, the disease appears in patched and initial symptoms are drooping and pallor of the leaves and later followed by yellowing and drying of the plant. Under moist conditions, a characteristic fan shaped mycelial strands of the fungus appear at the base of the infected plant. Later, brown mustard like round sclerotia develop on the mycelial growth.

Soil application of fungicides such as Brassical (0.1%), Bavistin (0.5-0.7%), Thiram (0.2%-0.3%) or Zineb (0.3%), three times at 20 days interval has been found to be effective in checking the disease. Certain cultural practices like reducing soil wetness, planting at wider spacing, destruction of infected debris and amending soil with organic matter are helpful in reducing the losses.

Flower Bud Rot [Erwinia sp.]

The disease appears mainly on young flower buds and results in dry rotting of the buds with brown scorched necrotic discoloration of peduncles. At later stage, buds shrivel and become dry.

Infected plant debris should be destroyed and burnt to check further infection. As the disease is known to be spread by thrips, remedial measures to control the thrips by applying suitable insecticide should be taken.

Botrytis Spots and Blight [Botrytis Elliptica]

Infected flowers show dark brown spots and ultimately the entire inflorescence dries up. Spots also appear on the leaves and stalks.

Spraying the plants with Carbendazim @ 2 g/l of water effectively controls the disease. The treatment should be repeated at 15 days interval.

Blossom Blight [Fusarium Equiseti]

Light brown lesions develop on petals, which soon darken and results in the drying up of the tissue. Infection is also noticed on flower stalk resulting in its collapse. Under humid conditions flower tips also become brown on which brown spore mass develop.

Spraying with Bavistin (0.02%) is useful in controlling the disease.

Alternaria Leaf Spot [Alternaria Polyanthi]

The disease manifests as brown spots with faint concentric rings on midrib and rarely on the margin of leaf. Occasionally peduncle may also be infected showing circular to oval spots which measure 10-30 mm in length and 4-5 mm in diameter. Infected leaves and peduncles become necrotic and dry up. Management is like as Botrytis leaf spot.

Diseases of Dahlia

Leaf Smut [Entyloma Dahlia]

The symptoms are first noticed on basal leaves during the month of June. Circular, light green spots appear on both sides of leaves. In severe cases development of numerous spots on leaves causes "shot hole" symptom and drying up of the leaves. Occasionally spots are observed on petiole and stem.

Dithane Z-78 (0.2%) or Dithane M-45 (0.2%) sprayed at 15 days interval during June-October is very effective for the control of the disease.

Powdery Mildew [Erysiphe Cichoracearum]

White to powdery patches or felt-like mass appears on the upper surface of the leaves. The disease is most prevalent in the early stages of plant growth and reappears at the end of flowering season.

Fortnightly sprays with Morestan (0.1%) Bavistin, Benlate (0.1%) or Wettable Sulphur (0.2%) is most effective to control the incidence of the disease.

Charcoal Rot [Rhizoctonia Bataticola]

The diseases is characterized by moist but firm rot of main stem and branches near the plant base. As the disease advances, progressive drying of the stem surface and shredding of stem becomes apparent. Black sclerotia are visible in the shredded tissue and this gives a smoky and charred appearance to the affected plants. In the final stages, the infected plants lodge and terminal twigs drop as a result of the breakdown of parenchymatous tissues.

Tuber treatment with Quintozene (PCNB) @ 0.5% for 10 minutes before planting has been reported to check the spread of the disease.

Stem Rot [Sclerotinia Sclerotiorum]

Infection appears on stem as light brown water- soaked patches that rot very quickly in cloudy weather particularly when coupled with shower. The rotten portion is covered with white frosty growth of mycelium on which black irregularly shaped sclerotia develop.

To control the disease heavy soil should be made porous with a mixture of sand and good drainage be provided. Plants should not be crowded. Affected plants should be uprooted and burnt. Drenching of soil with Quintozene (0.2%) is recommended to reduce the infection.

Wilt [Verticillium Alboatrum, Fusarium spp.]

These pathogens enter the roots, in which their presence is made evident by brown or black streaks along the conducting tissue. The plants wilt and die because of the blockage or destruction of the water conducting vessels, or death of the living tissues due to the toxins produced by the fungus. Since these wilt fungi may attack in storage, discolored or decayed parts should be cut away before planting.

Wilted plants should be destroyed and only healthy tubers should be used for propagation.

Crown Gall [Agrobacterium Tumefaciens]

Large tumors of abnormal tissue develop at the base of the plant and on the roots. Infected plants become stunted and the shoots spindly.

The disease can be prevented by dipping roots at planting in Streptomycin solution. Roots and crowns of plant with tumors should be destroyed and changing the location of planting will also help to curb the disease.

Bacterial Wilt [Ralstonia Solanacearum]

Plants usually droop and wilt suddenly. Stem near the soil rots and if cut, yellowish masses of bacteria ooze out.

The control measures are similar to those of crown gall disease.

Mosaic [Dahlia Mosaic Virus]

The leaves become mottled and develop pale green bands along the midribs and larger secondary veins. They are dwarfed and show a general mosaic or pale yellow spotting.

To prevent the disease, the roots of infected plants should not be used. Ruthless rouging of infected plants will often prevent the spread of the disease to healthy ones. Control of peach aphid (*Myzus persicae*), the vector of mosaic virus, will also help to prevent the disease.

Spotted Wilt [Tomato Spotted Wilt Virus]

It is also known as ring spot and appears as a characteristic ring pattern on the leaves.

Management: Spraying of 0.1% Malathion to control the vector insect is recommended. The disease can be checked by using disease free tubers or cuttings. Rouging of infected plants is helpful in minimizing the disease.

Diseases of Gerbera

Foot Rot [Phytophthora Cryptogea]

The symptoms appear as light brown water soaked patches on the collar portion of the stem or any other pant part which comes in contact with the soil. The leaves often turn yellow and the entire plant wilts.

Use well drained soil. Soil sterilization with Vapam at 100 ml/m^2 has been found to be very effective to control the disease. Drenching root zone of the plants with Metalaxyl (0.1%) can effectively minimizing the disease. Avoid overhead watering.

Root Rot [Rhizoctonia Solani, Pythium Irregulare]

The common symptoms of root rot are browning and rotting of the root system and purple discoloration of leaves. The infection results in stunted growth and ultimate death of the plant.

Soil sterilization and drenching with Copper oxychloride (0.4%) or Dithane M-45 (0.2%) gives good control of the disease.

Blight Or Grey Mould [Botrytis Cinerea]

Dead areas occur on leaves, flowers and stems of plants attacked by this fungus. The diseased leaf areas enlarge rapidly under favorable conditions and eventually entire leaves may turn black. The infected leaves and flowers become covered with a brownish gray mold.

All old leaves, flowers and diseased parts and plants should be removed. Plant should be spaced planted to improve air circulation. Spraying of Benlate or Bavistin (0.1%) controls the disease.

Sclerotium Rot [Sclerotium Rolfsii]

The initial symptoms are discoloration of the lower leaves and wilting of the young foliage commonly followed within a few days by the death and drying up of the entire plant.

Remove all the diseased plants. Treating the soil with Copper oxychloride and Dithane M-45 is effective in preventing the disease. Use well drained soil.

Powdery Mildew [Erysiphe Cichoracearum and Oidium Crysiphoides F. sp. Gerberae]

The pathogen covers the leaves with white powdery growth. In severe infection the vigorous leaves may be deformed and stunted.

The disease can be controlled by regular sprays of Wettable Sulphur or systemic fungicides like Benlate or Bavistin.

Leaf Spots [Phyllosticta Gerbericola Cercospora Gerberae, Alternaria Gerberae]

Leaf spots caused by P. gerbericola are mostly circular to irregular, brownish with dark brown margin. The fungus, Cercospora gerberae causes brown spots with darker margin on both surfaces of leaves but chiefly on the upper side. Alternaria gerberae cause circular to irregular, dark spots on leaves.

Spray plants with Bavistin (0.01%). Pick off and burn diseased foliage. The bio agents viz., Trichoderma viride, and T. hamatum have been found effective in checking the growth of the pathogen.

Ring Spot [Tobacco Rattle Virus]

Yellow or black annulated ring spot patterns on the foliage are produced by tobacco rattle virus.

Affected plants should be rouged out as soon as noticed. Soil heating before planting which helps to destroy the nematode vectors such as *Trichodorus* sp. is recommended for control of the disease. Aldicarb or Furadan is also effective to control the disease.

Diseases of Crossandra

Root and Foot Rot [Phytophthora Nicotianae]

The leaves show violet discoloration and dropping. In the advanced stage of infection wilting of whole plant can be noticed. The infection results in rotting of the roots and lower part of the stem and ultimately breakage of the stem at or near the soil level. Infected plants can be easily pulled from the soil.

The incidence of this disease can be greatly reduced by growing the plants on well drained sites. Prophylactic application of Captan (0.2%) as soil drench at the time of planting in the main field and application of Fosetyl-Al as soil drench 2-3 times at monthly interval are effective.

Fusarium Wilt [Fusarium Solani]

The foliage of affected plant first becomes dull yellowish green followed by wilting. The roots are more or less decayed and the base of the stem is brown at the soil level.

Infected plants should be pulled out and burnt. Apply Captan (0.2%) as a soil drench at the time of planting.

Stem Rot [Rhizoctonia Solani]

Tan brown to black lesion appear mainly on lower part of the stem. Young plants show rotting at the soil line.

Provide good drainage and avoid overcrowded planting. Remove and destroy the diseased plants.

Treating the soil with 1% Bordeaux mixture will be helpful in preventing stem rot.

Alternaria Leaf Spot [Alternaria Amaranthi var. Crossandrae]

Small, circular or irregular yellow spots appear on upper surface of the leaves. These spots enlarge and develop dark brown concentric rings in the necrotic tissues and induce yellowing symptoms.

Follow field sanitation. Spray plants with Mancozeb (0.2%) or Bavistin (0.1%) at 15 days interval during the growing season.

Leaf Blight [Colletotrichum Crossandrae]

Symptoms first appear on the lowermost leaves that have come in contact with the soil. Affected leaf consists of small, light brownish, depressed necrotic spots surrounded by reddish and slightly raised margins. On enlargement these spots become darker and the infected leaves soon become distorted and the edges tend to curl upward.

Maintain field sanitation, remove diseased debris and destroy it. Spraying the crop with Bavistin (0.2%) or Zineb (0.3%) is recommended.

Diseases of Lilium

Gray Mould [Botrytis Elliptica]

The disease is characterized by the presence of circular or oval and yellowish to reddish brown spots on the leaves. In some spots, central part is light gray in colour while the outer region is dark purple, shading into green healthy tissue. Flower buds may be shriveled, distorted, disfigured or may be killed depending on the severity of the attack.

The pathogen survives as sclerotia in the debris of infected plant parts that fall to the soil surface. These sclerotia are able to withstand adverse environmental conditions and microbial degradation till the next season when they produce spore inoculum that initiate new infection.

Use of disease free stock and resistant plant material such as Lilium gigantium, L. pyrenaicum, L. regale and L. willmottiae are recommended for healthy flower production. All the affected leaves and flowers of the plants should be collected and burnt. Spray with Bordeaux mixture @2% has been found useful. Fungicides like Captan, Thiram and Dithiocarbamate are also very effective. Foliar sprays of Carbendazim + Diethofenacol has given complete control of the disease while Cyproconazole and chlorothalonil reduced the disease severity to agreater extent.

Soft Bulb Rot [Rhizopus sp., Rhizopus Stolonifer]

Bulbs become soft and rotten and covered with a thick mat of mycelial growth.

Since the fungus enters only through wounds in the bulb scales, care should be taken that bulbs are not wounded at the time of digging and packing. Soil sterilization also prevents the disease.

Foot Rot [Phytophthora Cactorum]

It attacks stems just below the surface of the soil. Infected parts become shrunken and plants wither and die, bulbs are not damaged.

Plant only healthy, vigorous bulbs obtained from a reliable source. Good drainage is very important to prevent the disease and cultivation in wet weather is to be avoided. The affected plants should be dug up and destroyed. Bordeaux mixture or other copper fungicides should be applied to control the disease.

Bulb Rot [Fusarium Oxysporum F. sp. Lilii]

The initial symptoms are foliage yellowing and wilting. Although the bulbs may appear healthy, the roots develop a reddish coloured decay at their tips. The plants remain stunted with yellow foliage and with extensively rotted scales. The infected scales later fall away or shatter from the basal plate.

The fungus is soil borne and may be carried in or on bulb planting stock. It may survive in soil in the form of thick walled conidia or chlamydospores. The multiplication and spread of the disease is by production of new spores which are released into the soil. Inoculum dissemination can occur by soil movement, irrigation and through contaminated equipments from one field to another.

Removal of infected plants and drenching the soil with formalin has been found effective in minimizing the disease. Hot water treatment of bulbs for 2 hrs at 39 °C followed by a 30 min. dip in Benomyl eliminates infection and stimulates plant growth. Addition of mycorrhizal fungi can increase plant growth and offset the detrimental impact of root pathogens.

Brown Scale [Colletotrichum Lilii]

The outer scales become covered with small, circular, brown or black spots which spread rapidly in damp conditions to cause a soft black rot. The disease is bulb and soil borne.

Excessive watering and freezing of the bulbs should be avoided to prevent the disease.

Root Rot [Pythium Splendens]

Susceptibility to the disease is associated with viral diseases. Infected plants are severely stunted with lowest leaves dead or yellow with dead tips. Roots are either dead or reddish lesions appear on them.

Fungicides added as drenches give considerable control of the root rot. Planting should be done in well drained soil.

Bacterial Soft Rot [Pectobacterium Carotovorum]

The organism causes a soft, wet decay of bulbs. All affected bulbs should be discarded and healthy ones planted in a well drained soil. Care should be taken to avoid bulb injury during lifting from the soil.

Mosaic [Cucumber Mosaic Virus and Lily Mottle Virus]

The disease is characterized by irregular mottling and flecking on the leaves. Affected plants become dwarf and crippled.

Stunted and mottled plants should be collected and destroyed and bulbs from the diseased plants should not be used for propagation or planting. The disease is transmitted by aphids which should be controlled to check the disease. Spray Endosulfan to control the vector.

Rosette [Lily Rosette Virus]

Disease causes flat, rosette appearance of the plants, leaves being pale green or yellowish without mottling or streaking. The young leaves are curled downward and twisted. Bulbs from the infected plants become small, flat and tend to split.

Affected plant parts should be collected and destroyed and bulbs from the diseased plants should not be used for propagation or planting. To control the vector, use of insecticide spray is recommended.

Diseases of Anthurium

Anthracnose [Colletotrichum Gloeosporioides]

The disease primarily affects the individual flowers on the spadix. Infection starts as a tiny dark spots that expands later on. In advanced stages under wet conditions, a general rot of the entire spadix may occur.

Application of Mancozeb (0.2%) and Bavistin (0.1%) at 2 weeks interval can effectively control the disease.

Root Rot [Pythium Splendens]

Symptoms of the disease include reduced plant height, smaller leaves and flowers and a general lack of vigour. In severe cases all the roots may be rotted.

Planting should be done in well drained soil. The disease can be controlled by drenching the soil with Carbendazim (0.1%).

Leaf Spot [Septoria Anthurii]

The symptoms appear on leaf as yellowish spots which later become dark brown and black. Severe infection may result in defoliation of leaves.

Disease can be controlled by spraying Zineb (0.3%) repeated at interval of 2-3 weeks.

Powdery Mildew [Erysiphe communis]

The pathogen covers the upper leaf surface with white powdery growth. Powdery growth also appears on the stem.

Treatment with 0.1% Benomyl easily controls the disease.

Bacterial Blight [Xanthomonas Axenopodis pv. Dieffenbachiae]

Initial symptoms of the disease include irregular shaped water soaked spots surrounded by slight yellowing on lower surface of leaves. In advanced stages spots become darkened and are encircled by yellow zone. Systemic infection is characterized by blackening of the stem and leaf sheath.

Strict sanitation measures, the removal of affected leaves and avoidance of close planting are recommended. Three to four spraying with Streptomycin sulphate (0.02%) at 10-12 days interval should be done. First spray should be given as soon as the disease starts appearing.

Diseases of Narcissus

Basal Rot [Fusarium Oxysporum F. sp. Narcissi]

The disease first causes decay in the root plate or at the base of the scales which in warm weather develops during later stages of growth in the field although basically it is a disease of narcissus bulbs during storage or transit. The tops of the infected bulbs turn yellow and die before normal time of maturity. At digging time the affected bulbs may show many purplish roots which start dying from the tip towards the base. The affected bulbs at planting result in premature yellowing of the foliage.

The pathogen survives in soil and in infected bulbs. Soil temperature below and near 13 °C causes less infection but increases rapidly with increase in temperature upto 29 °C coupled with sufficient moisture.

Soil fumigation before planting with Methyl bromide (50g/m²) or Metham sodium (metham, 75 or 100 ml/m²) or during growing season two applications of Aldicarb (10% a. c. ; 4g/m²) effectively control basal rot disease. Spraying the plants with 0.5% Zineb, 0.2% Ferbam, 0.5% Captan or 1-1.5% TMTD which together control even *Botrytis narcissicola*. Preplanting bulb treatment and spray application with *Streptomyces* sp. has bean found effective in disease management.

Smoulder [Sclerotinia Narcissicola]

It attacks the leaves mainly at ground level and produces yellow, malformed and stunted shoots, and the plant rots. Secondary infection occurs on damaged leaves and sheaths and flowers can also be badly spotted. Bulbs infected in storage may rot and even slight infection of the bulb may develop inoculum in the field on planting causing rotting of bulbs in the field before sprouting or producing yellow and weak shoot which die afterwards.

The affected bulbs and the plants should immediately be collected and destroyed. As the Inoculum continues to develop year after year through the infected bulbs present in the soil, rotation should be followed. Repeated rouging of plants. Fungicidal bulb dips and fungicidal sprays just before flower picking and from leaf maturity to senescence is recommended for controlling this disease. Benomyl and Carbendazim (0.1%) sprays applied 14-20 days before and after picking controlled the disease effectively.

Green Mould [Trichoderma Viride]

Its infection ultimately results in rotting of the bulbs. The pathogen invades the bulbs through the wounds and its appearance is realized by bright green growth.

Care should be taken not to bruise or injure the bulbs. Bruised bulbs should be discarded at the time of either grading or when putting them in stores. The bulbs which have been found affected with the disease either at lifting or after storage should be discarded and other bulbs be treated with Captan.

Penicillium Bulb Rot [Penicillium Corymbiferum]

The pathogen causes rotting of bulbs in narcissus cv. Rembrandt and Carlton. Disease can be controlled by the spray of Captafal (0.2%).

White Root Rot [Rosellinia sp.]

Affected bulbs show black rotting generally of outer scales with white fungal strands in the form of wooly mass near the base plate, especially under moist condition.

Affected bulbs should be destroyed and infected fields avoided for planting of narcissus. Use of Captan (0.2%) may control this disease.

Whit Mould or Ramularia Blight [Ramularia Vallisumbrosae]

Small sunken grey or yellow spots appear on leaves and green parts. Later spot enlarge and darken to yellow brown with yellow margin. Moist warm weather accelerates the spread of this disease by producing the spores in the form of a mass of grey or white powder on leaves which rapidly spread by wind or in water. In cool dry weather the spread of the disease is checked. This disease infects and destroys the foliage, and in late flowering cultivars it spoils the flowers by attacking the flower stems.

Do not replant for one year in fields where the disease has occurred. The pathogen may be controlled by three or four sprayings with 4:4:40 Bordeaux mixture. Spraying with Mancozeb (0.2%) is also useful in checking the disease.

Leaf Scorch [Stagonospora Curtisii]

This pathogen causes dark brown lesions and spots starting from the neck of the bulb to the leaf tip at emergence. These brown spots spread to the rest of the leaf and flower stems and flowers during wet weather and make a burnt appearance. With the growth of the foliage, these brown blisters burst and discharge the spores which travel to healthy foliage and plants through rains or moist weather. The premature death of foliage prior to several weeks of its natural death also adversely affects the bulb growth. The spores of this fungus remain around the neck of the bulbs after the death of the foliage.

Hot water treatment of bulbs is useful in disease control. Bulb disinfection with Granosan, Formalin, Captan, TMTD or Cycloheximide and spraying with 1% Bordeaux mixture or Mancozeb (0.2%) to control the disease is recommended.

Fire [Botryotinia (Sclerotinia) Polyblastis]

Disease symptoms appear as pale brown lesions which quickly spread on the leaves. On the perianth of the flowers small pale brown spots appear at the packing time. Under moist conditions, the spread of disease is enhanced and the flowers are first killed and then leaves are infected where it over winters.

The disease can be eradicated by destroying the plant debris and by removal of the infected flowers. Bordeaux mixture at 4:4:40 and tank mix Zineb may control this disease if sprayed regularly at 15 day intervals.

Narcissus Yellow Stripe [Narcissus Yellow Stripe Virus]

The plants infected by NYSV, even at emergence from the ground show bright yellow stripes as streaks which may be seen up to flowering time in the field but not during flowering. The infected plants produce less number of flowers and poor quality blooms, the number and weight of bulbs are also reduced.

Infected plants should be eliminated. To control the aphid vector, insecticide spray is recommended.

Diseases of Stock

Damping Off [Pythium Debaryanum, Rhizoctonia Solani, Pellicularia Filamentosa]

Basal parts including roots of the infected plants turn black and die. When collar region is invaded, the seedlings topple down and then die away.

Disinfection of seeds and soil sterilization reduce the disease losses. Drenching of infected soil with Carbendazim (0.1%) or Copper oxychloride (0.2%) is effective in checking the disease.

Grey Mould [Botrytis Cinerea]

Under humid conditions the flowers of mature plants are covered with a dense growth of grey mould releasing a large number of light grayish spores.

Sterilization of seeds and spraying of seedlings with Carbendazim (0.05%) or Mancozeb (0.2%) have been recommended to check the disease spread.

Wilt [Fusarium Oxysporum F. sp. Matthioli and Verticillium Alboatrum]

Wilting of seedlings and older plants which is characterized by yellowing, premature leaf drop, stunting, vascular discoloration and wilting.

Planting should be avoided in wilt infested soil. Fumigation of infested soil with Chloropicrin is suggested to eradicate the soil borne fungus.

Leaf Spot [Alternaria Raphani]

The disease is characterized by formation of round spots on leaves, stem and flowers, 1.5-2.0 cm in diameter. The spot on close examination reveals the formation of concentric rings and exhibits profuse blackish sporulation under moist conditions.

Collect and destroy infected crop debris to reduce the initial inoculum. Spraying with fungicides like Dithane M-45, Dithane Z-78 or Blitox at 0.2% is recommended to check the disease.

Downy Mildew [Peronospora Parasitica]

The infected leaves develop pale green spots on the upper side. The lower side of the spot shows downy mould growth at the point of infection. The attacked stem and flower parts do not develop normally and remain dwarf.

Spraying with Dithane M-45 (0.2%) at regular intervals checks the disease spread. Spraying seedling every 3-4 days with mixture containing 40% Copper oxychloride and 41% Zineb proved to be the best control measure of the disease.

White Rust [Albugo Candida]

On the under surface of lower leaves white raised pustules develop. The disease spreads on the filed by white powdery spores released form these pustules. Later the lesions become yellowish to brown in colour.

For managing disease effectively remove diseased plant and destroy the plant debris. Spraying of Copper oxychloride (0.2%) is recommended to control the disease.

Viral Diseases

Stocks are infected by several viruses eg., mosaic, curly top, wilt and flower breaking. The methods of eliminating these diseases include rouging out and destruction of plants showing symptoms and control of leaf hoppers by using Malathion or Endosulfan.

References

- "Orchid Research Program". Michigan State University. Archived from the original on 22 September 2012. Retrieved 17 October 2012

- Diseases-of-Ornamental-Plants-and-their-Management: researchgate.net, Retrieved 7 March, 2019

- Verchot-Lubicz, J.; Ye, C.-M.; Bamunusinghe, D. (1 June 2007). "Molecular biology of potexviruses: recent advances". Journal of General Virology. 88 (6): 1643–1655. Doi:10.1099/vir.0.82667-0. PMID 17485523

- Inouye, N (2008). Viruses of orchids: symptoms, diagnosis, spread and control. Netherlands: Blue Bird Publishers. P. 176. ISBN 9789079598038

- Pearson, M. N.; Cole, J. S. (1 November 1986). "The Effects of Cymbidium Mosaic Virus and Odontoglossum Ringspot Virus on the Growth of Cymbidium Orchids". Journal of Phytopathology. 117 (3): 193–197. Doi:10.1111/j.1439-0434.1986.tb00934.x

- Gubler, W. D.; Davis, U. C.; Koike, S. T. "Powdery Mildew on Ornamentals". University of California- Agriculture and Natural Resources. University of California Statewide Integrated Pest Management Program. Retrieved 3 December 2015

Permissions

Index

* 9 7 8 1 6 4 1 1 6 5 3 1 0 *